Enzymes

Thomas W. Traut

Enzymes

The Worker Bees of the Cell

 Springer

Thomas W. Traut
Chapel Hill, NC, USA

ISBN 978-3-031-93607-4   ISBN 978-3-031-93608-1 (eBook)
https://doi.org/10.1007/978-3-031-93608-1

© The Editor(s) (if applicable) and The Author(s), under exclusive license to Springer Nature Switzerland AG 2025

This work is subject to copyright. All rights are solely and exclusively licensed by the Publisher, whether the whole or part of the material is concerned, specifically the rights of translation, reprinting, reuse of illustrations, recitation, broadcasting, reproduction on microfilms or in any other physical way, and transmission or information storage and retrieval, electronic adaptation, computer software, or by similar or dissimilar methodology now known or hereafter developed.
The use of general descriptive names, registered names, trademarks, service marks, etc. in this publication does not imply, even in the absence of a specific statement, that such names are exempt from the relevant protective laws and regulations and therefore free for general use.
The publisher, the authors and the editors are safe to assume that the advice and information in this book are believed to be true and accurate at the date of publication. Neither the publisher nor the authors or the editors give a warranty, expressed or implied, with respect to the material contained herein or for any errors or omissions that may have been made. The publisher remains neutral with regard to jurisdictional claims in published maps and institutional affiliations.

This Springer imprint is published by the registered company Springer Nature Switzerland AG
The registered company address is: Gewerbestrasse 11, 6330 Cham, Switzerland

If disposing of this product, please recycle the paper.

To Karyn Traut

Prologue

When speaking about science or research, a common question is "What is it for?" Most people understandably are focused on the practical.

Research has two general purposes: First is the exploration of areas of knowledge with the aim of understanding them and, in so doing, discover something new. Second is applying that newly found knowledge to enable medical advances, household conveniences, and in general, make life more livable for humans.

In a beehive, the queen and drones are needed to produce fertilized eggs that will become future bees. But the worker bees perform all the varied and necessary functions for the colony to be able to continue. In a similar fashion, enzymes are responsible for all the biochemical reactions that occur in a living cell. While all worker bees are generalists, who can each perform the same set of varied duties, enzymes are specialists that normally bind only one, or a few chemical compounds, and perform some form of special chemistry to produce the many molecules necessary for the maintenance of life.

This book summarizes mountains of research by many scientists in the field of general biochemistry, with an emphasis on enzymes, which make all the important molecules in our body, such as DNA, proteins, etc. Enzymes are also essential for removing all toxic chemicals that inadvertently enter our body. All metabolic reactions in our body are performed by enzymes.

But perhaps, most importantly, it should give the reader a satisfying feeling of knowing: "Aha! That is how my cells work."

Acknowledgment

I am indebted to Charles Lewis Jr. and to Karyn Traut for their thoughtful suggestions. The illustration in Fig. 11.1 was provided by Jason Traut.

Competing Interests The author has no competing interests to declare that are relevant to the content of this manuscript.

Contents

1 **Conditions Necessary for Life: Take Minerals, Water, Oxygen, and Stir**   1
  1 What Are Enzymes? Meet the Worker Bees   2
    1.1 How Were Enzymes Discovered?   4
    1.2 pH and Buffers   6
    1.3 The Affinity of Enzymes for Their Substrates   7
    1.4 Inhibitors   10
    1.5 How Are Enzymes Named?   11
    1.6 How Fast Are Enzymes?   13
    1.7 How Many Enzymes Are There?   14
  Resources   16

2 **What Are These Things Doing in My Body? Overview of Living Cells**   17
  1 Constituents of a Normal Cell   18
  2 Cell Size Corresponds to the Cell's Function   20
  Resources   23

3 **Our Superbees: How Fast Are Enzymes?**   25
  1 How Strongly Do Enzymes Bind Substrates?   29
  2 Biochemical Molecules Are Generally Fairly Stable   30
  3 Evolution Optimizes Biochemistry   33
  Resources   35

Contents

4 How Did We Get Here? Starting Conditions for Life on Earth — 37
 1 The Importance of the Early Elements — 39
 2 The Importance of Oxygen — 42
 3 Abiotic Precursor Building Blocks — 44
 Resources — 46

5 The Nuts and Bolts of Metabolism: How Cells Function — 47
 1 Overview of Metabolism — 47
 2 A Brief Look at Energy Formation — 48
 3 Metabolism Recycles Molecular Compounds — 50
 4 The Importance of Thermodynamics and Equilibria — 51
 5 Is there an Energetic Reason for the Common Molecules in Life? — 55
 6 How Is Metabolism Organized? — 58
 Resources — 59

6 The ABCS of Protein Structure and Synthesis — 61
 1 Types of Protein Structure — 63
 2 Proteins Vary in Size and Complexity — 68
 3 How Information Is Positioned on a New Protein — 71
 4 Bonds that Hold Proteins in a Stable Conformation — 74
 5 Synthesis of Proteins — 75
 Resources — 77

7 Our Amazing Hemoglobin: Blood and the Circulatory System — 79
 1 We Have Many Globin Proteins — 82
 2 Having Different p50 Values Is Important for Globin Functions — 87
 3 Oxygen Binding Curves — 87
 4 Fetal Hemoglobin Binds Oxygen More Tightly — 90
 Resources — 92

8 Allosteric Enzymes: Shape Shifters — 93
 1 Hemoglobin Was the First Allosteric Protein Observed — 94
 2 There Are Two Types of Allosteric Proteins — 97
 3 The Nature of V-Type Enzymes — 98
 4 G-Proteins — 103
 Resources — 104

9	**Sex and Nucleic Acid Enzymes**	**105**
	1 The History of Molecular Biology	107
	2 Replication of DNA	111
	3 Many Origins of Replication Are Required for Timely Duplication	114
	4 Overview of Molecular Biology	116
	5 Other Control Regions in a Gene	119
	6 Processing RNA Transcripts	119
	7 Improvements in DNA Sequencing	121
	8 The Importance of Electrophoresis	122
	9 Procedure for DNA Fingerprinting	124
	10 Mutations and DNA Repair	126
	10.1 Autosomal Cell	127
	10.2 Reproductive Cell (Sperm or Oocyte)	127
	10.3 Potential Mutations	130
	11 Regulation of DNA Expression	134
	Resources	136
10	**Uh, Oh, My Genes Are Changing: Evolution and Diseases**	**137**
	1 Recombination of DNA Segments by *Nucleases* and *Ligases*	138
	2 Antibody Diversity	139
	3 Racial Differences Are Only Skin Deep	142
	4 Tyrosinases Are Isozymes	143
	5 Vitamin D Synthesis Occurs in Skin Cells	147
	6 Medical Radiation Jargon Explained	149
	7 Sickle Cell Anemia: A Helpful Deleterious Mutation?	151
	8 Treatments for Sickle Cell Anemia	155
	Resources	157
11	**Control Your Sugar: Special Enzymes Control Glucose Metabolism**	**159**
	1 Overview: The Importance of Glucose	160
	2 Other Sources of Glucose	166
	3 Glycolysis: The Pathway for Producing the Initial ATP	167
	4 Key Regulatory Enzymes in Glycolysis	172
	5 Starvation	176
	Resources	180

Contents

12 Enemies Within? Our Own Enzymes Assist Viruses in Infecting Us — 181
   1 How Do Viruses Invade Us? — 182
   2 Why Are There Viruses? What Are They Good For? — 184
   3 Common Influenza Virus — 187
   4 HIV-AIDS — 188
   5 The Resistance Allele — 188
   6 SARS-CoV-2 or Covid-19 Virus — 191
   7 How Bad Is Covid? — 194
   Resources — 195

13 Traitors: Cancer Results When Regulatory Proteins Become Mutated — 197
   1 The Cell Cycle and Cancer — 200
   2 Tumor Suppressors — 203
   3 A Mutated Rb Tumor Suppressor Leads to Retinoblastoma — 204
   4 p53 Tumor Suppressor — 206
   5 Oncogenes — 208
   6 Viruses Can Cause Cancer — 209
   Resources — 210

Epilogue — 213

Index — 215

1

Conditions Necessary for Life: Take Minerals, Water, Oxygen, and Stir

Abstract A description of living organisms, using simple bacterial and eukaryotic cells as examples. The cell's metabolism is described by analogy to an automobile factory, where different components of the final automobile are made on their own assembly lines before merging to achieve the final complete car. Enzymes are compared to the individual workers who handle specific tasks along such assembly lines. Enzymes are then more specifically described as having an active site, or catalytic pocket, that has evolved to be the ideal size and shape for a substrate to bind there, so that it can be chemically altered. A brief history of enzymology is included. The importance of a cell's pH and the concept of enzyme affinity are then explained, which is summarized as the enzyme's K_M, and illustrated by a Michaelis–Menten kinetic plot. Enzyme nomenclature is explained, and enzyme rates (speeds) are defined.

Keywords Enzyme · Catalytic site · Substrate · Inhibitor · Affinity · Buffer

What is life? A philosopher might have to think about this, as there have been many attempts to answer what would appear to be a simple question. Most answers agree that living organisms are complexes of molecules that have been organized to perform multiple functions. Among these is the ability to maintain their own existence by procuring energy, in the form of food for animals or sunlight for plants. A living organism must also be able to reproduce itself, either by itself or with the help of a mate. We will soon see that viruses cannot do this unless they infect a cell that acts as a host to provide the molecular machinery needed for the virus to make copies of itself.

Living organisms respond to external stimuli, normally in an appropriate fashion for the organism's well-being. Living beings communicate with members of their own species, normally for the benefit of the group. Even bacteria do this by exchanging DNA segments with each other. If a bacterial cell experiences a mutation in one of its genes, it might be saved if the DNA segment that it receives contains a normal version of that gene.

In the age of artificial intelligence, it may now become possible for robots to do all the above, including replication and communication. Only philosophers may then be able to define what life is.

The simplest forms of life are single-celled organisms, among which are found all the bacteria, many types of algae, amoebas, etc. Such cells contain enough DNA to code for all the proteins that are needed for the many chemical reactions occurring in such a cell. These reactions include the generation of energy that a cell needs for many of its functions, as well as the synthesis of many large molecules from simple chemical precursors. We generally define this complex of thousands of reactions as the cell's metabolism.

1 What Are Enzymes? Meet the Worker Bees

The workings of a factory can be a simple analogy for describing cellular metabolism. Consider an automobile factory, where multiple assembly lines continue to produce separate components of an automobile, such as the car body, doors, windshields, etc. Each such assembly line involves multiple steps, where smaller components used in the manufacture of the final product are attached in the proper sequence until the car body is completed. In the twenty-first century, much of this assembly work is now automated, but in the early years of car manufacture, it was people who became efficient at doing a very specific step correctly and fairly rapidly.

Enzymes are the workers in the cell's metabolism, which is the umbrella term for all the reactions occurring in our cells. The synthesis of all larger molecules occurs in a sequence of reactions, often called a pathway, where each step in such an assembly is performed by a specific enzyme. Should an enzyme for a particular chemical step be in limited supply, or even completely absent, the cell will be unable to make enough of the desired product, and in most cases, the cell will not survive. This makes evident how specific most enzymes are. With humans in a factory, if Joe is absent, Paul or Juan could fill in for him. At first, they might not be as fast as Joe was, not having done Joe's job before, but they could start at a moderate pace and soon be as good as Joe had been. Enzymes are not this versatile.

1 Conditions Necessary for Life: Take Minerals, Water, Oxygen...

To continue the assembly line analogy, if Joe's specific function was to add the outer panel to a car door and then tighten the nuts onto bolts that would hold the door panel on, Joe could just as easily attach a fender and tighten the nuts to the bolts that hold the fender to the car's frame. In this example, positioning and tightening the nuts would be comparable to a single chemical reaction. With enzymes, that would be the only chemical step that a particular type of enzyme could accomplish. The other part of Joe's assembly job involves the door panel, and Joe is good at handling this, positioning it correctly with the bolts in the right places and then attaching nuts to the bolts.

But, unlike Joe, few enzymes are even flexible enough to do a similar reaction with a different component, such as a fender in this analogy. With an enzyme, the chemical component involved in the chemical reaction is defined as a *substrate*, normally designated by **S** (see Fig. 1.1). As shown in this figure, the substrate is the name for the chemical molecule or compound that the enzyme can bind to and chemically alter.

Because Joe has hands and can think, he can easily transfer his normal activity of adding door panels to a different but similar job and attach a car fender. Somewhat comparable to Joe's hands, an enzyme has an active site for binding the component to be added in making a molecule. But enzyme active sites are not as flexible as human hands, and therefore each enzyme has an active site that has been selected by evolution to be the right shape to bind just one or two very special compounds. As depicted in Fig. 1.1, enzymes normally have at least three amino acids and/or cofactors positioned at the

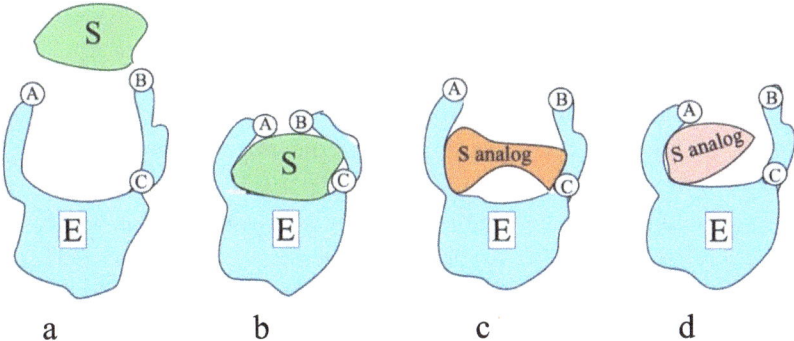

Fig. 1.1 An enzyme with an active site for forming an E-S complex. (**a**) The substrate (S) approaches the active site. (**b**) Substrate binding induces the proper alignment of three amino acid or cofactor groups on the enzyme: A, B, and C. (**c**) and (**d**) Substrate analogs (competitive inhibitors) bind to the enzyme, but the catalytic groups are not aligned properly and the analog binds more weakly

catalytic site that bind to the incoming substrate and orient it so that the chemical reaction can occur at the correct position on the substrate.

Most enzymes are so fast in their reactions that they can be considered super catalysts, and, like a cartoon Superman, they should be depicted with a bold **E**. They do only one chemical step, but they do it very efficiently (making very few mistakes) and normally very fast. That is why nature has evolved a single unique enzyme for almost every chemical step in cellular metabolism.

What is a *catalyst*? The word was coined from the Greek for dissolving or dissolution. In its original meaning, it is an agent that causes something to come apart. The earliest enzymes observed were hydrolases (Greek for splitting with water), and by splitting a larger molecule, they caused it to come apart or dissolve.

1.1 How Were Enzymes Discovered?

Early in the nineteenth century, attention to the process by which starch was converted into a simple sugar, or the process by which cane or beet sugar (sucrose) could be fermented to alcohol, increased with the discovery that the *polarimeter* could measure such chemical compounds by the change produced in their polarization of light, as more molecules were produced in an experiment. The polarimeter had just been invented by French scientists in the early 1800s, and it measures the rotation of light. In some chemical reactions, a newly made compound in a solution can be measured by its optical rotation, which made it possible to determine the increase of such a compound over time. It was therefore possible to quantify the formation of products in a chemical reaction, and thereby begin proper studies of the factors that influenced these systems.

In studies almost 200 years ago, the original term *enzyme* was implied by two French scientists who were curious about how yeast cells could chemically change starch into the common sugar molecule glucose. Starch is a polymer (from Greek: many parts), meaning it is a long chain composed of some type of simple building blocks or monomers, in this case, glucose. In the early 1800s, there was debate as to how this chemical change was performed. Because it was the question of how fermentation was accomplished by yeast cells that the active agent, whatever it might be, was named a *ferment*.

Were intact yeast cells necessary, or was it perhaps done by some component, a *ferment*, within the cells? The two scientists, Payen and Persoz, broke up the yeast cells, so as to release the interior contents, and then tested this extract for its ability to produce glucose. They observed that an extract from the yeast cells could accomplish the chemical conversion of starch into glucose. Anselme Payen, with a degree in chemistry from the École Polytechnique, had become the manager of a commercial business that prepared and sold various chemicals. Although remarkable, it is also quite understandable that, in this very first paper dealing with enzymes, the title of the paper referred to the "application to the industrial arts" for such enzyme preparations. They reported that this chemical activity occurred *en zyme*, meaning "in yeast," where "zyme" was borrowed from the Greek word for leaven, because yeast is a leavening agent. Their work was soon followed by German and English scientists, who simply joined these two words into the now standard *enzyme*.

In this original publication, the French scientists worked with an enzyme that cleaved starch molecules, and they used the Greek term for cleavage, *diastasis* (meaning "to separate"), but made it more properly French with an "ase" ending: *diastase*. While Payen and Persoz may not have intended for "-ase" to become the standard suffix in enzyme names, subsequent work by scientists in Table 1.1, who continued such research, led to new enzyme names almost always ending in "-ase."

There are only a few exceptions to this naming rule among metabolic enzymes. These are the digestive proteases discovered at almost the same time, such as *pepsin* (discovered in 1836 by the German physician, Theodore Swann). Swann may not have seen the paper by Payen and Persoz in a French journal, and so he named his enzyme *pepsin*, from the Greek word for digestion. *Trypsin* and *chymotrypsin* are two other digestive proteases that do not follow standard nomenclature.

Table 1.1 Important dates in the emergence of enzymology

Date	Event
1833	Payen and Persoz describe a soluble extract from yeast cells that converts starch to the simple sugar glucose; they name the active agent *diastase*, thereby beginning to name enzymes with an "ase" suffix. They also emphasized that this activity occurred "en zyme," French for "in yeast."
1836	Theodore Swann isolates *pepsin* from the stomach
1876	Willi Kühne suggests the term *enzyme* for these catalytic activities that were previously called *ferments*
1894	Wilhelm Ostwald defines a *catalyst* as increasing a chemical reaction, while itself remaining unchanged
1902	Victor Henri defines the "enzyme-substrate complex" and also an equation for enzyme activity as a function of the available concentration of its substrate
1913	Michaelis and Menten present an equation that defines an enzyme's rate as a function of the substrate concentration and also of how strongly the enzyme binds to S

It required almost a century of additional work to confirm that enzymes were proteins. In recent years, it has been established that there are a very few chemical reactions that can be performed by RNA molecules, named *ribozymes*, but because more than 99% of known enzymes are proteins, in this book, we will generally consider enzymes to be proteins.

As more of these early studies of enzymes were being done, it became desirable to describe the enzyme's activity in standard terms or units that could then be measured, so that different scientists could compare different enzymes as to how fast they were and how strongly or weakly they bound to their substrates. The first to do this was the French biologist Victor Henri, who proposed that the enzyme and substrate first formed an E-S complex, as shown in Fig. 1.1, before any chemical change could occur, and also proposed that for each enzyme there would be a maximum or highest activity, which today is normally written as V_{max}, where V represents the velocity or rate of the reaction, and the subscript "max" stands for maximum.

1.2   pH and Buffers

When we speak about pH, we are referring to how acidic a solution is. The term and definition were proposed by a Danish scientist, Søren Sørensen, and it is an abbreviation for the German "potenz Hydrogen" which means the power (or concentration) of hydrogen ions. The range for pH values goes from 0 to 14, where 0 means the solution is highly acidic, 7 means that it is neutral, and 14 means that it is very alkaline (also called "basic"). Mammals normally have cells and body fluids with a pH very near 7.4.

Human or mammalian metabolism has a tendency to produce acidic molecules, which could therefore lower the pH of a cell or of the blood. To prevent excess acidity in the stomach after a meal, people use common buffering tablets known as antacids, because they prevent or minimize acidity. These are normally made with compounds such as calcium carbonate or magnesium hydroxide.

Modern enzymology became more standardized when a Canadian scientist, Maude Menten, worked with a German professor, Leonor Michaelis, and observed that the acidity of a solution influenced an enzyme reaction. They introduced the use of buffered solutions for studying enzymes, which would prevent any meaningful change in acidity during the time period of an enzyme experiment. Readers will appreciate that the amount of time for such experiments is important, because such measurements almost always measure how much product is formed per minute or per hour. If the enzyme's rate is very slow, it may take many hours before enough product is formed to be measurable. During this extended time, the enzyme molecules must remain stable, or they might become less active, and that is why buffered solutions became necessary.

1.3 The Affinity of Enzymes for Their Substrates

Michaelis and Menten also realized that there was a second important feature that determined enzyme activity: how well (or how tightly) does the enzyme bind its substrate? This is now normally described as the *affinity* of the enzyme for its substrate, and for each different enzyme, this should then be a constant feature. For Michaelis, a German, this word was *Konstante*, and therefore represented by the letter K or k in an equation. They formulated an equation that is still commonly identified with their names: the Michaelis–Menten equation:

$$\frac{v}{V_{max}} = \frac{[S]}{K_M + [S]} \tag{1.1}$$

In the equation above, "v" represents the measured *velocity* or rate, under any condition of the experiment, because both the time for measuring activity and the amount of available substrate [S] could be altered. The brackets for [S] denote substrate concentration. Vmax has already been defined, and the new constant, K_M[1], denotes the affinity of the enzyme for the substrate, S. K_M and S are measured in units of substrate concentration. These units are standardly expressed in molarity, which is defined as 1 mole[1] of some compound per liter of solution or water. Chemical compounds generally vary in their actual atomic weight, just as people do in body weight. For example, H_2O has an atomic weight of 18 grams, CO_2 of 44 grams, and glucose of 180 grams. But a molar solution of each of these contains exactly the same number of molecules.

To help understand this, consider a room with 30 5-year-olds and another room with 30 adults. These two types of people have very different body weights, but each room would have the same concentration of people.

One mole contains an enormous number of molecules, so it is routine to do experiments with more dilute solutions, where 1/1000 is a millimole, and 1/1,000,000 is a micromole. Solutions with such concentrations are then designated as mM (millimolar) or μM (micromolar).

---

[1] One mole is defined by Avogadro's number, which is 6 × 10^{23} molecules.

It should then be evident that when the concentration of S is high enough to equal the K_M, then v will be equal to one-half of V_{max}:

$$\frac{v}{V_{max}} = \frac{[S]}{K_m + [S]} \text{ Then if } K_M = [S], \frac{[S]}{[S]+[S]} = \frac{1}{2} \quad (1.2)$$

Thereafter, it was easy to measure the K_M for an enzyme. Visual inspection of Fig. 1.2 shows that the velocity curve at first rises and then begins to approach a plateau.

In the early years of using this method, scientists simply estimated how much higher V_{max} might be, as shown by the horizontal dashed line. While a calculation of a K_M from an experiment, as shown in Fig. 1.2, is not as precise as one would like, it is fairly close to the true value and also easy to determine experimentally.

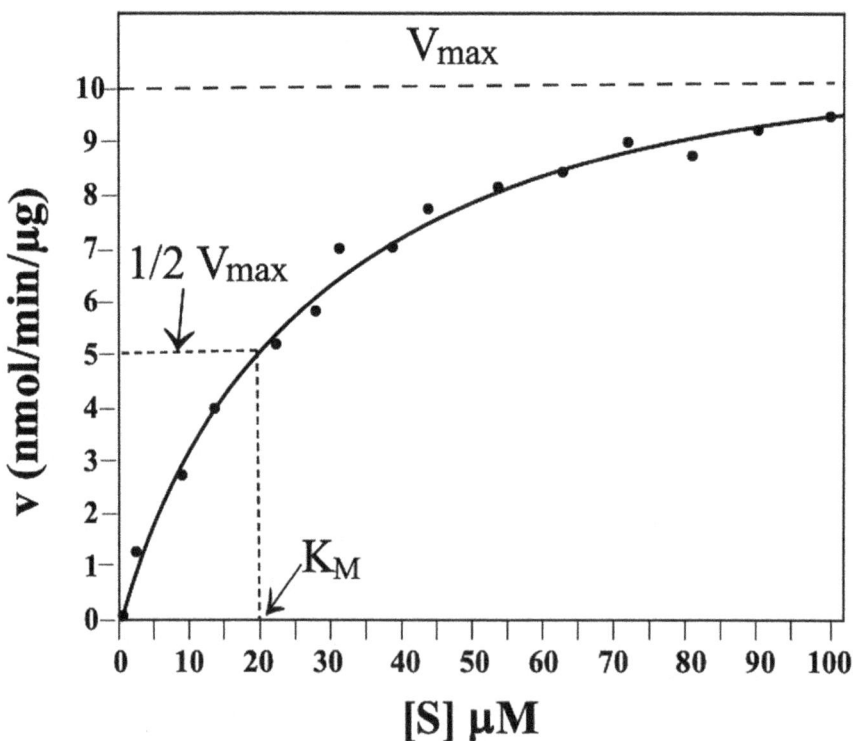

Fig. 1.2 This Michaelis–Menten plot fits the data points and sets V_{max} at 10 nmol/min/μg protein. The measured K_M is then 20 μM

1 Conditions Necessary for Life: Take Minerals, Water, Oxygen… 9

An interesting historical anomaly: although almost every biochemist would call Fig. 1.2 a "Michaelis–Menten" plot, the truth is that in their famous paper, Michaelis and Menten never presented a graph such as shown in Fig. 1.2. They used a mathematically different approach that showed a graph that looked very different from Fig. 1.2, but it produced a value for K_M. The familiar graph with their name was created some years later by other scientists, but it honored the original authors.

To understand concepts such as affinity, V_{max}, and K_M, I will use as an example an anecdote from my time as a high school senior working at a summer job in a cannery. This was in California. The company, not far from my home, processed and packaged vegetables and fruits arriving continuously from nearby farms and orchards.

For two weeks, I worked on the assembly line for packing fancy, choice apples. I knew that these were "fancy, choice" apples because those words were on the cardboard boxes into which I was packing the apples.

To justify this description, the apples were first washed in a large tub. A truck with apples would back up to the tub, which was about 5 feet high and 8 feet in diameter, and raise the truck bed so that the apples would all tumble into the tub, which was half full with water. The tub contained large brushes on metal rods that were rotated by a motor so as to move the apples around and remove any dirt that might be on them. The water was then drained and fresh water added to rinse the apples again. A conveyor belt on one side of the tub, and at an angle going to the top of the tub, had raised ridges across the belt at intervals, and these would hold apples as they were moved up and out of the tub. While moving along this passage, hot air was blown onto the apples so that they would be dry before they landed on the main conveyor line belt.

This conveyor belt was about 2 feet wide and 20 feet long. Young men, such as myself, were positioned about 3 to 4 feet apart, and at my position, I had a stack of empty cardboard boxes and a small stand with boxes of green tissue paper into which apples would be packaged individually before going into the cardboard box. The procedure for packing apples was simple: with my left hand grab a piece of tissue paper and hold it flat in that hand, and with the right hand, reach over the moving conveyor belt to grab an apple. Place the apple onto the paper, pull up the corners of the paper, and twist these around enough to enclose the apple. Then put the apple into the box, in which there was a receiving cardboard tray, much like the lower half of an egg carton but with depressions much larger to be suitable for holding an apple.

For the first 5 minutes, perhaps, I definitely concentrated on each action that was needed, but this was a very simple procedure, and so I was soon on autopilot in continuing. I was young and would be going to the big university at the end of summer, and so had many things to contemplate about my new life there. The speed, or *velocity*, at which I packed apples would be measured by the number of apples I packed per hour. My maximum v (V_{max}) would occur when the conveyor belt was completely full with apples. Then, no matter where I positioned my right arm, I would grab an apple. In this example, I was acting mechanically, like a robot, exactly the way enzymes behave. Enzymes are like miniature robots; they have one function, and they do that automatically. In the sense of working like an enzyme, and because the apples were my "substrate," I could have been called an "apple packerase."

Sometimes the trucks with new apples would be slow, or some other problem could occur, and the abundance of apples on the conveyor belt would become much smaller. If I merely reached my arm out without looking, I would not grab an apple each time, so my rate v would go down. But I needed to keep my summer job to earn enough money for my new independent lifestyle in the fall, so I would become alert when the apple supply became diminished and work at almost full speed. Enzymes do not see or think, so if the concentration of their substrate is lower, then they will produce less product.

It is also useful to appreciate that enzymes and their substrates exist inside cells, in a fluid matrix, where both enzyme molecules and substrate molecules float and drift. The enzyme does not reach out to grab a substrate; rather, the substrate encounters the enzyme by bumping into it, and if it should bump into the catalytic pocket (see Fig. 1.1), it will interact with the residues at the active site and remain momentarily while some catalytic action is performed on the substrate. It should be evident that an enzyme's affinity for the substrate (i.e. how well it binds S) is a function of how appropriately the enzyme's active site is shaped for this binding step, and also of how many S molecules are available to reach the enzyme's active site.

1.4 Inhibitors

Evolution has helped to select that, for each enzyme, its active site pocket is the right size and shape for its normal substrate. As diagrammed (Fig. 1.1), any other molecule that has a size and shape somewhat similar to the normal substrate can also bind at the catalytic site, and while it resides there, it prevents the normal substrate from entering. In most cases, the enzyme can perform its chemical function on the inhibitor molecule. Why then is this called inhibition? Because v for an enzyme is defined by how much of the desired normal product it makes for the cell per minute, or per second if the enzyme is fast. While an enzyme is performing catalysis on an inhibitor molecule, it is not making anything that the cell needs; therefore, it is less productive, and this is inhibition.

To help make this clearer, while I was an apple packerase, full-time workers had heard about my college plans and also observed how I sometimes worked on autopilot. As a joke, to test the college kid about how smart he was, one man came near to where I was stationed and placed an old scruffy baseball on the conveyor belt near me. I automatically grabbed the baseball without noticing any difference; after all, it was the right size and shape. I apparently wrapped it in the green paper and placed it into the cardboard box. I was unaware of doing this, but about 30 min later, when it was time for our morning break and I went into the resting area, men began to point and jeer at me. The smart college kid could not even distinguish a baseball from an apple. They were sure that I would be lucky to last even one week at school in the fall. In this anecdote, the baseball is a competitive inhibitor, which I then processed as if it were a substrate. It is "competitive" because the baseball competes with the apples to be in my hand, and it momentarily inhibited me from packing the desired apple, which is what the grocer wants when he opens the cardboard box.

1.5   How Are Enzymes Named?

Perhaps as early as 10,000 years ago, after the last ice age receded and simple farming and agricultural advances were originated, peoples in many different areas discovered that if either grain or fruit were left in containers and became soaked with water and burst open, an interesting new element appeared in the fluid within such containers, and drinking this tended to make people more mellow and happier enough that they everywhere chose to do this in a regular fashion.

This was the discovery of fermentation, by which grains can easily be the source for beer, and fruits—mostly grapes and berries—the source for wines. This new interesting element is alcohol, although that identification only occurred thousands of years later. The process for converting starting materials containing sugar to alcohol required the addition of yeast, and this soon became described as fermentation. Therefore, the active agents in yeast cells that actually performed this transformation were originally called *ferments*.

Most enzymes bind two substrates, as shown by this general scheme: A + B $\rightleftarrows$ C + D. One substrate is often a general molecule containing a group that is widely used in biosynthesis, such as ATP, which is the most important molecule for providing chemical energy for thousands of metabolic reactions. The structure is shown in Fig. 1.3. Such a general substrate is therefore the donor substrate. The second substrate is a specific acceptor molecule to which the transferred group will be added, in this example, the phosphate group from ATP.

Enzymes are then frequently named for the acceptor substrate, and for the type of transfer shown below (Table 1.2). In group 1, *glucokinase* is named for the acceptor substrate glucose and for its action in transferring a phosphate group onto glucose.

When this activity was first observed, this step was considered to be "kinetically activating" glucose for subsequent metabolic reactions, and since then, most enzymes performing phosphotransfer reactions are named *kinases*.

The second enzyme in Table 1.2 removes a hydrogen atom from glyceraldehyde-3-P, or "dehydrogenates" it.

The fourth example, *nucleotidase*, is a hydrolase, meaning that it uses a water molecule to transfer the -OH from the H_2O to replace the bond between two nucleotides in a nucleic acid. Because "hydro" or "hydra" is often a prefix for words dealing with water, and *lysis* is the Greek word for splitting, such

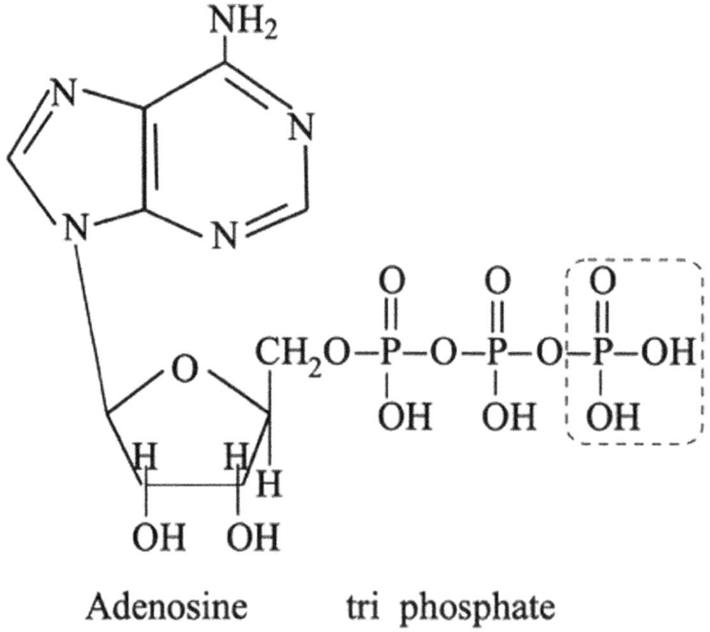

Fig. 1.3 Structure of Adenosine TriPhosphate (ATP). The terminal phosphate group at right, in the dotted box, is transferred by *kinases*. Abbreviations for atomic elements: *H* hydrogen, *C* carbon, *N* nitrogen, *O* oxygen, *P* phosphorous

Table 1.2 Enzyme nomenclature

1	Named for: substrate	+	(catalytic action)	+	ase
	examples: Gluco (se)		kin		ase
	glyceraldehyde-3-*P*		dehydrogen		ase
	DNA		methyltransfer		ase
	nucleotid				ase
2	Named for a gene - examples: src, ras, onc				
3	Named for size of protein - examples: *p*21, *p*53, *gp*42, *pp*60 (here *p* stands for protein, *gp* for glycoprotein, *pp* for phosphoprotein)				

reactions are hydrolytic, and enzymes performing these reactions are *hydrolases*. It became standard to interpret an enzymatic name without a specified function as a hydrolase.

Enzymes in group 2 were discovered more recently, and they were not named for their function, which was still unknown when they were first observed, but were named for the distinctive features of the cell or tissue in which a gene coding for this protein was identified as being associated with some disorder or disease.

To make labeling of these newly identified genes easy, they were customarily named by a three-letter code, normally an abbreviation for the clinical feature. Therefore, *src* refers to sarcoma, a type of cancer; *ras* for rat sarcoma, where this gene was first observed—it is a GTPase[2] involved in activating specific enzymes (Chap. 8); *onc* for oncogenesis, as this gene was identified with the onset of cancer (Chap. 13). It took years to establish that these enzymes are involved in the regulation of when cells begin to duplicate, the necessary step for forming tumors.

The third group above includes enzymes first identified by techniques for determining their size (indicated by the number value in the name, representing the protein's molecular weight in kilodaltons), and also if they had any chemical modification, such as the addition of a phosphate or a sugar group. This type of nomenclature has become very widespread, and there are thousands of enzymes with such cryptic names.

1.6 How Fast Are Enzymes?

All chemical reactions necessary for life are sufficiently slow that one or more unique enzyme catalysts are required to accelerate the reaction and make the needed product almost immediately available. For cells to survive, natural

---

[2] GTP = guanosine tri phosphate, one of the five classes of nucleotides.

selection has provided that each enzyme is always fast enough, with the slowest metabolic enzymes having a rate of at least one catalytic reaction per enzyme molecule per second. This is normally defined as the enzyme having a rate $\geq 1\ s^{-1}$, where "$\geq$" means "equal to or greater than," and "s^{-1}" means "per second." Therefore, all normal reactions in cellular metabolism must occur in a time of one second or faster. These units define enzymes as very fast—even the slowest enzymes.

We are used to units for speed, especially with autos or airplanes, of miles per hour. An hour contains 3600 seconds, so a car at a highway speed of 60 mph is moving at one mile per minute, or 88 feet s^{-1}. While a rate of $1\ s^{-1}$ is the slowest observed for metabolic enzymes, most enzymes are much faster, with rates of 100–500 reactions per second, and some are amazingly fast, with rates of about one million reactions per second!

1.7 How Many Enzymes Are There?

There is no simple answer for this, as it depends on which organism is considered and varies by how complicated the organism is. At the lowest end is the bacterium *Mycoplasma genitalium*, a urogenital pathogen (a simple combining of urine plus genital regions). The normal version of this bacterium has 482 genes coding for proteins and small RNAs, of which almost all are enzymes. More common bacteria, such as the *E. coli* found in the human gut, have about 4200 genes, most of which would code for enzymes.

The central metabolism of mammalian or human cells is not very different from that of common bacteria. And in bacteria, central metabolism requires almost 4000 enzymes. Because humans have so many different types of cells, with discrete physiological and biochemical properties, there are over 22,000 genes in total, and perhaps 18,000–19,000 of these code for proteins. At least three types of different proteins make up this greater number:

1. Structural proteins: Such as actin, myosin, troponin, collagen, etc.
2. Isozymes: Slightly different versions of the same enzyme that perform the same chemical reaction but have some different features to make them perform appropriately in different tissues or organs. Some examples are listed in Table 1.3.
3. Regulatory proteins: Necessary for controlling the process of development of the embryo and to make the adult organism more flexible in responding to environmental opportunities or stresses.

Table 1.3 Protein functions

I. Ligand binding:		
	a. carriers:	1. hemoglobin (O_2)
		2. transferrin (Fe^{++})
		3. albumin (fatty acids)
	b. receptors:	1. membrane-bound: for hormones, etc.
		2. immunoglobulins (for antigens)
		3. enzyme regulatory subunits
	c. enzymes:	like receptors/carriers, but they chemically modify the ligand
II. Structural:		collagen, histones, actin, myosin, tubulin, etc.
III. Signaling:		1. peptide hormones
		2. growth factors (i.e. transcription factors)

It is difficult to obtain a meaningful answer to the question of how many enzymes there are in humans. Searches on the Internet yield multiple websites that claim there are 75,000 different enzymes—with no description of how this number is obtained. At the lower end are websites stating that there are only 1300 enzymes in humans. This answer is clearly too low or too simple—depending on how the authors defined enzymes, which was not explained.

In 2024, the website BRENDA (**BR**aunschweig **EN**zyme **DA**tabase) listed 8423 different enzymes from all organisms. So, about 5000 appears reasonable for humans.

In this book, we will focus mostly on enzymes. But, as shown in Table 1.3, proteins have many important functions, and enzymes are the most abundant type, and also have many interesting properties and functions. However, we will consider a few examples of some other proteins, such as hemoglobins (oxygen transporters; from the Greek haima = blood, and the Latin globule = small globe or sphere), transcription factors (involved in gene regulation and cancer), and so on.

Simple bacterial cells have almost no duplicated genes. But gene duplication is quite normal in human cells. This may reflect what humans call a cost-benefit analysis, though this is done somewhat automatically by evolution. It costs a lot of ATP to synthesize a duplicate copy of some gene. ATP (Fig. 1.3) is the most important molecule for providing the energy needed for synthetic reactions. In Table 1.4, we see that gene duplication is sometimes remarkable. Modern DNA sequencing has made it normal to identify duplicated copies of any gene. It is then interesting to observe how extensively some genes have been duplicated, when it appears that not all of the extra copies function to produce some protein variant by the duplicated genes. Here, duplicated genes produce *isoproteins* (Greek *iso* = same); they are not normally exactly the same, as evolution via mutations normally causes moderate changes in their function.

Table 1.4 Examples of isoproteins

Protein	# genes	# diff. proteins
Collagen	31	18
Carbonic anhydrase	20	15
(Hemo)globin	13	10
Hexokinase	12	4
Lactate dehydrogenase	15	3

Nevertheless, for the common structural protein collagen, which serves to make the framework for all bones, tendons, cartilage, etc., humans have actually evolved 18 slightly different types of collagen, while 13 of the duplicated genes appear to have no function. Hemoglobin is also an interesting example. Most people know that hemoglobin is the oxygen carrier in our red blood cells. But in fact, we have two different types of adult hemoglobin, and as fetuses we did not use the adult hemoglobin genes, but instead used fetal hemoglobins (explained in Chap. 7), and muscle has a different version known as myoglobin.

If humans have about 22,000 genes, and only 18,000–19,000 of these code for proteins, what are the remaining genes for? To become useful, all genes are initially transcribed into RNA (**ri**bo**n**ucleic **a**cid) molecules. RNA is a nucleic acid very similar to DNA (**d**eoxyribo**n**ucleic **a**cid), so this process may be likened to making a photocopy of some piece of paper or document. When genes code for a protein, their newly made messenger RNA (mRNA) will be used as a template for making the appropriate protein. But the cell also needs many RNAs that function as they are, most frequently in the process of making new proteins (described in Chap. 6). These genes that do not code for proteins then code for the many RNA molecules that the cell also needs.

Resources

Discussion of biocatalysts (enzymes): https://en.wikipedia.org/wiki/Enzyme
Brief review of enzyme kinetics and types of graphs for analyzing them: https://en.wikipedia.org/wiki/Enzyme_kinetics
A translation of the original German publication that made the study of enzyme kinetics more straightforward: *The Original Michaelis Constant: Translation of the 1913 Michaelis–Menten Paper.*
Kenneth A. Johnson and Roger S. Goody https://pubs.acs.org/doi/10.1021/bi201284u

2

What Are These Things Doing in My Body? Overview of Living Cells

Abstract Comparison of cell sizes with the cells' functions, and emphasis on the small sizes of most cells, which are more easily described in metric units for size, followed by a comparison of American units for measurements with metric units. Life began more than 4 billion years ago with simple bacterial cells, such as *E. coli*. Eukaryotic cells are more complex, having more organelles plus a separate nucleus to protect the cell's DNA.

Keywords Cell sizes · Metric units · First bacterial cells · Eukaryotes · Cell nucleus · Chromosomes · Cytoplasm · Mitochondrion · Viruses

Most people can remember high school biology classes, where cells could be seen with a microscope. These cells were certainly very small and were therefore invisible to the human eye without some special viewing device. With only a few special exceptions, living cells are microscopic in size. This size range largely reflects how large a cell must be to contain the various components that are necessary for the cell to function normally. Initially, with the earliest cells that came into existence almost 4 billion years ago, the length of such a bacterial cell was about 1–2 microns. The American standard units of length have no sizes small enough to describe a bacterial cell, but the metric system can do this exceedingly well. In the twenty-first century, there are 195 different nations, and only 3 of them do not use the metric system: the United States, Liberia, and Myanmar.

Table 2.1 shows values for very small sizes in metric units and their corresponding values in American units for length. An additional symbol, the

Table 2.1 Units for measuring size

Unit	Size (cm)	Size (mm)	Size (microns)
Inch	2.54	~25	~25,000
Foot	~30	~300	~300,000

squiggle (~), will also be frequently used to indicate that some number is approximately this value. As a simple example, an article about a sports match might state that there were 987 spectators. This number implies an exact count, perhaps as shown by the number of tickets sold. But for convenience, we might simply round this number up and say there were about one thousand (~1000) spectators.

To appreciate the metric scale, and how small cells are, note that a small unit of length for everyday purposes, the inch, is about as long as 25,000 microns. In metric units, one *E. coli* cell, a common bacterial species found in the human gut, is about 1.5 to 2 microns long. It would take about 12,000 to 15,000 *E. coli* cells, end to end, to equal one inch.

1 Constituents of a Normal Cell

The bulk of a bacterial cell is the cytoplasm, Latin "cyto" for cell plus "plasm" for "something formed." Here (Fig. 2.1) is where all the metabolic activity of such a simple cell occurs. The many molecules that form metabolic pathways are too small to be visible at this scale. The exception is the cell's DNA or RNA, which are sufficiently dense and long to be visible under a microscope. There are also numerous small organelles, ribosomes, which work constantly to make new proteins, as described later in Chap. 6.

Enzymes, the proteins that perform most of the biochemical reactions, are also too small to be visible at this scale. Other features evident in Fig. 2.1 are the cell wall enclosing the cytoplasm and various thin filaments attached to the outer wall. These include the pili, which act as contact sensors, and the flagella, from the Latin for "whip," which can be agitated and move the cell through any liquid environment and can turn the cells as directed by contact via the pili.

These pili have multiple functions, including contact between mating cells, so that *E. coli* cells may pair up and exchange some of their DNA with each other, by which they minimize mutations occurring in their DNA when they acquire an undamaged copy from another cell.

2 What Are These Things Doing in My Body? Overview of Living Cells

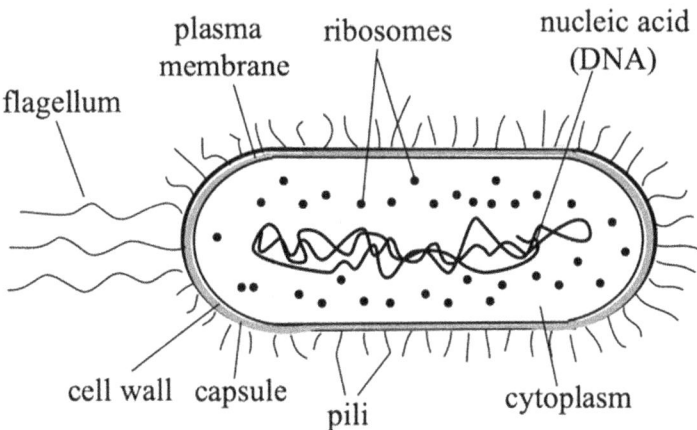

Fig. 2.1 Diagram of an *E. coli* bacterial cell, about 2 microns (μm) long

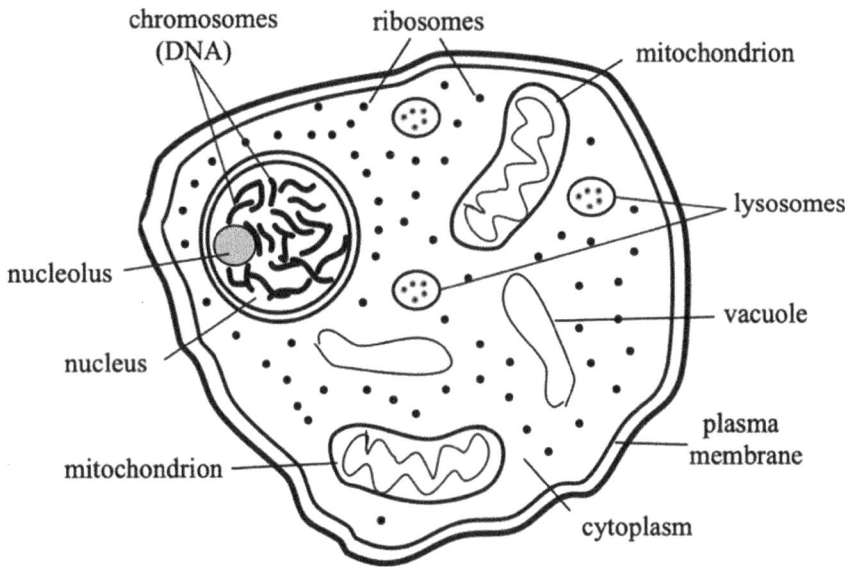

Fig. 2.2 A Typical eukaryotic mammalian cell. Note that membranes separate organelles from the cytoplasm, and also surround the cell. Mammalian cells are normally at least ten times as large as a bacterial cell. **A typical virus would be this big → °**

Eukaryotic cells, from the Greek "eu" for good and "karyon" for kernel, meaning that they have a separate and defined nucleus, are more complex (see Fig. 2.2). It is apparent that cells contain many visible organelles within the cytoplasm. An important feature, and for which these cells are named, is the nucleus, a separate organelle with its own membrane to protect the DNA within from potential damage by *nucleases*: enzymes necessary for dismantling the cell's damaged DNA or foreign DNA in an infectious agent (described in Chap. 9).

Contained within the nucleus is the nucleolus, where *ribosomes*, which are complexes containing four RNA strands and 33 separate proteins, are continuously produced. The synthesis of cellular proteins is performed by these ribosomes, and because protein synthesis is a constant cellular requirement, new ribosomes need to be made at an equally steady rate. Unlike bacteria, which normally have a single DNA strand, eukaryotic cells have a larger amount of DNA, which is always separated into many strands known as *chromosomes* (from Latin for "colored body"), as shown in Fig. 2.2. Human cells have 46 chromosomes, with one set of 23 chromosomes coming from each parent.

Other important organelles include the *mitochondria* (plural for mitochondrion; from Greek *mitos* "thread," + *khondrion* "little granule"), essential for producing ATP, the key energy molecule of the cell, discussed in Chap. 11. There are also *lysosomes* (Greek for loosening bodies) that contain hydrolytic enzymes that degrade unneeded proteins and other polymers (Greek: many bodies) and have an acidic interior similar to the mammalian stomach. These mini garbage disposals are essential. There are human diseases resulting from the poor functioning or depletion of lysosomes.

An additional class of organelles are the vacuoles, small vesicles (from the Latin vacuola for "empty"). They only look empty under the microscope, and depending on the type of cells in which they are located, vacuoles can have many different functions, including storage of water, storage of degraded molecules ready for expulsion from the cell, etc.

2 Cell Size Corresponds to the Cell's Function

While most bacterial cells are very small, eukaryotic cells have evolved to become more complex and to have more diversified functions. By becoming more specialized, they can also form a multicellular organism, of which animals are good examples. We are familiar with the many types of organs and tissues in a mammal such as a human, and we can therefore appreciate that different parts of our body, having different functions, may also need cells of different sizes, with some examples in Table 2.2.

Platelets are the smallest cells in a human. They occur in the blood and are required to prevent the loss of blood after an injury. Their function is to

2 What Are These Things Doing in My Body? Overview of Living Cells

Table 2.2 Sizes of human cells

Cell type	Size (microns)
Platelet	2–3
Erythrocyte (red blood cell)	7–8
Skin cell	30
Ovum	120
	Size (meters)
Spinal nerve's axon	~1

aggregate, in response to an injury signal, forming a large complex of these tiny cells that can then lodge in and plug the injury where blood is leaking out. Platelets can be this small because they do not have any of the organelles shown in Fig. 2.2. Thus, they are simply very small sacs enclosed by a membrane, but with special receptor proteins on the outside of their membrane that permit them to sense the injury signal and which then also permit them to be part of the blood-clotting cascade, described in Chap. 8.

Erythrocytes (from Greek erythros = red, plus kytos = basket) are the red blood cells that carry oxygen from the lungs to all the tissues. They are also unusual, again having no nucleus, which may use up 50% of the interior volume of a cell. After having eliminated these organelles, this additional space within each cell permits it to contain large quantities of hemoglobin, the protein that binds oxygen and releases it when red blood cells pass through capillaries in any tissue. The size of a red blood cell is therefore determined by the internal diameter of the smallest capillaries through which blood must flow. As a red blood cell slides through a capillary, and the cell is the same diameter as the inside of the capillary, the membrane of the red blood cell slides along the membrane of the surrounding capillary so that oxygen diffuses directly through these membranes into the adjacent cells.

Skin cells form a protective layer on the outside of the body and therefore can be larger. Their size is limited by the fact that when injury occurs, the skin cells immediately adjacent to that site will die. If skin cells were too large, then too large a surface area on the body would need to heal by making new cells, even when the injury is quite small.

Some human nerves have an unusual size. Here, it is not the main cell that is very large, but the axons of these spinal nerves that must extend for about 1 meter along the full length of the spinal column to reach a muscle or a sensory cell.

Eggs, comparable to a human ovum, are so much larger (Table 2.3). Because almost everyone is familiar with chicken eggs, one can then see how an egg

Table 2.3 Eggs—the largest single cells

Species	Length of egg	Weight of egg
Elephant bird	16 inches	16 lbs
Emu	12 inches	3 lbs
Emperor penguin	4 inches	4 lbs
Ostrich	6 inches	5 lb

Table 2.4 Virus genome sizes

Species	Size of RNA	Strands
COVID (SARS-CoV-2)	29.9 kb	1
HIV (human immune-deficiency virus)	9.8 kb	2
Influenza-B	14.5 kb	8
Size of DNA		
Human cell	3.1 Gb	2 × 23

becomes larger as the bird laying this egg becomes larger. Larger eggs are not required in large birds because they have more DNA, but because a greater amount of yolk, the main food source for the developing chick, is needed as the developing chick inside an egg needs a longer time to reach the point where it can emerge. For larger birds, the egg also needs more albumin (egg white), which serves as an initial supply of amino acids to make proteins required for the chick's growth.

These features then enable the hatching chick to be larger when it emerges from the shell so that within minutes it can move, walk, and forage for worms, seeds, etc. If the chick is destined to become an ostrich, it must start as a much larger chick. The egg of the elephant bird, long extinct, was as large as an American football. At the smallest size of biological organisms are the viruses.

In Table 2.4, sizes are not in metric units, as viruses are too small even for these units. Here, size is defined by the amount of genetic material required for making such a virus. Both RNA and DNA are formed by linking nucleotides, with these basic units forming a long strand of a nucleic acid. When two DNA strands match up with each other to form the double helix, the hydrogen bonds holding these strands together are formed between the bases that are always a part of a nucleotide (detailed in Chap. 9).

The size of any nucleic acid is defined by how big it is, i.e. how many bases it contains. Most readers are familiar with kilograms or kilometers, where "kilo" represents 1000. The value for HIV in Table 2.4 of 9.8 kb abbreviates the term kilobase to "kb," meaning that its RNA contains 9800 bases. Humans are much more complicated than a simple virus, with instructions for making almost 20,000 different proteins, while HIV codes for only 15 viral proteins.

A nucleic acid 1000 times larger than this virus would have dimensions in megabases, where "mega" signifies one million. The total size of the human genome contains 3.1 billion nucleotides, defined as 3.1 Gb, where G represents "giga" for billion.

Resources

Structure of a bacterial cell: https://micro.magnet.fsu.edu/cells/bacteria-cell.html

Structure of a mammalian cell: https://www.britannica.com/science/cell-biology

Comparison of US and metric systems of units: https://www.about.ch/various/unit_conversion.html

3

Our Superbees: How Fast Are Enzymes?

Abstract Discussion of enzyme rates, with the observation that enzymes protecting a cell against chemical or radiation damage must be faster than the spontaneous rates for such damage. This establishes a lower rate for enzyme activities. Exponential numbers are explained. An explanation of why molecular switches and restriction enzymes are slower than metabolic enzymes. An explanation of an enzyme's *affinity* for its substrate = how strongly it binds.

Keywords Enzyme rates • k_{cat} and k_{non} • Chemical damage • Radiation • Radicals • Enzyme affinity • K_M • $t_{1/2}$

We have a natural sense for time frames that define human actions in the larger physical world. Due to enthusiasm for sports, many people have an idea of what the fastest rate is for running, cycling, or swimming. We are not as interested in lower limits. We will again see a range of rates, depending on the actual task to be performed. Because of our interest in sports or in any effort, we expect that any timed process should achieve the fastest possible rate. Evolution is less demanding. For a species to survive, each of its enzymes must be just good enough or fast enough to accomplish what the cell needs.

A critical starting point for this question is the normal rate of damage to a living cell by the various ever-present sources of chemical and radiation damage, even though these are normally at low levels. Examples include damage from oxygen radicals that form spontaneously, from various types of chemical damage, and from radiation damage largely due to ultraviolet light rays.

There is a variety of possible oxygen radicals. Figure 3.1 shows an example with a hydroxide radical, where the oxygen would prefer to bond with 2 H^+ atoms to form H_2O (water). The second H^+ can dissociate, forming the hydroxide ion. If this molecule is struck by a high-energy photon, one electron can become dislodged, forming the radical, which is unstable, as the hydroxide is energetically more stable by having the missing electron. Replacing this electron can be accomplished by making a bond with any available compound, thereby chemically altering it.

Damage from reactive radicals may occur to almost any molecule in a cell, but has long-term results mainly when it involves DNA. Therefore, at least a subset of enzymes, with the responsibility of preventing such damaging agents or of reversing their effects, must have reaction times that are faster than the natural rates for damage, examples of which are in Table 3.1. The third column in this table shows the chemical (nonenzymatic) rate at which damage occurs naturally. The last column shows the catalytic rate for enzymes that protect against this damage.

To understand the rates shown in this table, one must understand the use of exponential numbers. For each numerical value, the superscript after the number "10" means that 10 should be multiplied by itself that many times. For example, the formation rate for metabolic acidity is 10^2, or 10 × 10, which is 100; 10^3 is 1000, 10^4 is 10,000, and 10^6 is one million.

We see in Table 3.1 that oxygen radicals occur at a rate of 10^4 s^{-1}, meaning 10,000 such toxic oxygen molecules are produced within a cell each second. And also that the enzyme *superoxide dismutase* has a k_{cat} of 10^4 s^{-1}, meaning that the enzyme can undo 10,000 oxygen radicals per second. Therefore, this enzyme is just fast enough to prevent any oxygen radicals from existing for more than a second.

The use of these exponential numbers became necessary when scientists in certain areas began to require such large numbers to describe their studies. Astrophysicists describe the age of the universe as about 14 × 10^9 years old, or 14 billion years. Geologists look at rock layers that are 10^8 years old, or 100 million years old. And biochemists who study enzyme kinetics deal with an even wider range of possible times. Then, in order to display such time scales on a graph, it would be impossible to have large numbers with too many

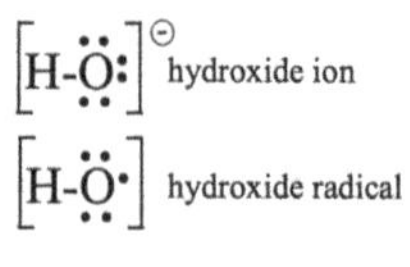

$[H\text{-}\ddot{\underset{..}{O}}:]^{\ominus}$ hydroxide ion

$[H\text{-}\ddot{\underset{..}{O}}\cdot]$ hydroxide radical

Fig. 3.1 Hydroxide ion and hydroxide radical. The dots are electrons

Table 3.1 Natural sources for chemical damage and protective enzymes

Source	Intracellular agent	Formation rate (s^{-1})	Protective enzyme	k_{cat} (s^{-1})
Oxygen	Oxygen radical (O_2^-)	10^4	Superoxide dismutase	1×10^4
	Hydrogen peroxide	$\geq 10^4$	Catalase	1×10^6
Metabolism	Acidity	$\geq 10^2$	Carbonic anhydrase	1×10^6
UV light	High-energy photon	0.1	Photolyase	0.4

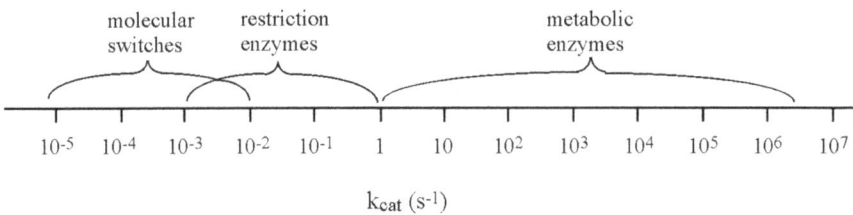

Fig. 3.2 Catalytic rates for enzymes related to their function. Metabolic enzymes perform all chemical steps for the synthesis and interconversion of cellular molecules. Restriction enzymes cleave DNA; molecular switches activate critical enzymes

zeroes. An extreme example is Avogadro's number, defining how many molecules are in one mole: 6×10^{23}. This number in the normal format is:

600,000,000,000,000,000,000,000.

Also evident in Fig. 3.2 are negative exponential numbers. These are the reciprocal values for the positive exponentials. We now know that 10^2 equals 100. Then 10^{-2} equals 1 divided by 10^2 or 0.01. To make this more evident: an enzyme with a k_{cat} of 10^2 s^{-1} performs its chemical function 100 times in one second. If k_{cat} is 10^{-2}, then it takes the enzyme 100 seconds for a single chemical step. This is much slower.

We see in Fig. 3.2 that metabolic enzymes are always faster than the regulatory enzymes. That is possible because metabolic enzymes are interconverting chemical molecules within the cell. Speed is desirable so that they can produce many molecules of product per second. These enzymes may occasionally make a mistake in their chemical function by binding an incorrect substrate and then making an unwanted product. But as long as they make enough of the needed product, and if the incorrect product causes no harm, then these occasional errors are acceptable. Here, speed is clearly more desired than accuracy.

In contrast, *restriction enzymes* cut DNA at very specific positions (Fig. 3.3). These enzymes were named when they were originally discovered because where they cut DNA appeared "to be restricted" to exact sequences along the DNA. All bacteria have restriction enzymes. These enzymes are used as a defense by each species of bacteria against invading DNA from any other species. These enzymes are useful because they cut at DNA sequences that do not exist in that particular bacterial species. So—should they be infected with a virus or plasmid—the host bacterial cell's restriction enzymes should cut up the foreign DNA, but not their own, preventing infection and damage.

If the enzyme is not accurate and lines up with the cell's own DNA, the result may lead to damage to the cell's genome, and this usually causes some impairment or cell death. Therefore, in order for restriction enzymes to avoid making mistakes, they act more slowly so that they can position themselves very exactly along the correct DNA strand that they will then cut. It should also be noted that here, speed is not that important. Invading DNA strands only need to be cut once or twice. This is not a process that must be done constantly. Therefore, accuracy is more important than speed.

Molecular switches are enzymes that are activated by a chemical modification, and then they, in turn, initiate some temporary physiological process. An example is the release of sugar from the liver when a sudden need for rapid muscular activity occurs. This process is activated by hormones such as adrenaline, which is produced when the brain detects some critical need for rapid action or movement. Because these enzymes are slower, there are enough of them to perform their needed task so that enough of their target enzymes, *glycogen phosphorylase* in this example, with adrenaline and the release of sugar, are activated.

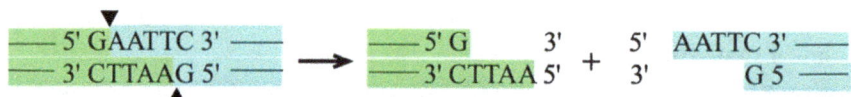

Fig. 3.3 A double-stranded DNA with the exact sequence at which the bacterial restriction enzyme *EcoR1* binds and cuts. The horizontal lines on the 5′ and 3′ ends indicate the continuation of the DNA strand in both directions. ▼ and ▲ indicate where the enzyme cuts the DNA. *EcoR1* is a dimer with two identical catalytic centers that align on opposite DNA strands with the exact same DNA sequence, 5′ to 3′, and therefore cut at the exact same site on the opposite DNA strands

1 How Strongly Do Enzymes Bind Substrates?

The strength of enzymes for binding their substrate has already been defined as the K_M, described in Chap. 1. The actual values for most metabolic enzymes are in the low millimolar or high micromolar range. Evolution tuned each enzyme to have a K_M that is very similar to the normal concentration of that metabolite in the cell, or in some body fluid. If one thinks about this, it means that metabolic enzymes will be busy at almost all times and will be working at a rate that is about one half of V_{max} (see Fig. 1.2). Then, should the concentration of that substrate increase, after a meal, for example, the enzymes still have the ability to increase their rate.

Readers must understand that the cytoplasm has a limit for how many compounds may be dissolved in it at any time. Therefore, free amino acids, sugars, or nucleotides are never kept at large concentrations. If they are not used for synthesizing the polymer into which they can be formed, they are recycled for storage as fat or for energy. Table 3.2 shows concentrations for the most common metabolites, which then also suggest the affinities that enzymes using these metabolites will have.

An additional feature influences the enzyme's affinity for the substrate (how tightly it binds). In Chap. 1, it was described how binding is a constant on/off process, but because the enzyme makes weak temporary bonds with the substrate, the off-rate is slower than the on-rate. But, as binding gets tighter, the off-rate gets slower, both for the substrate and also for the product being made, which will have many of the bonds with the enzyme that the enzyme made with the substrate. So, as binding of the substrate becomes stronger/tighter, the rate will become slower. And, depending on how much of the product is needed per minute, this may require some other compensation, such as making more of these enzymes. A team of ten workers, with the same ability, can do twice as much work as a team of five.

An exception to the concentrations in Table 3.2 is glucose. As will be described in Chap. 11, glucose is the sugar that is mainly used in providing ATP for most energy processes. Therefore, in blood, the normal concentration of glucose in a healthy adult is about 5 mM. If any readers are diabetic, you will have been informed that this value is 90 mg/pct and that you should

Table 3.2 Concentrations of major metabolites

Important macromolecule	Normal precursor	Concentration (mM)
Protein	Amino acids	0.1–3
Nucleic acids	Nucleotides	0.1–3
Carbohydrates	Simple sugars	0.1–2.5

make an effort to keep your blood glucose at 130 mg/pct or lower. "90 mg/pct" actually means 90 mg of glucose sugar in 100 ml (milliliters) of blood, or whatever other fluid is being considered.

These apparently different values are for identical concentrations of glucose. They are different because biochemists use the international system of units, and American hospitals and clinics still use the older system from pharmacologists. Glucose has a molecular weight of 180 grams (usually "g"). Then a 1 molar solution (1 M) would contain 180 g/L or 18 g for one tenth L (= 100 mL, or pct). A 1 mM solution would then be 18 g/pct. Then the biochemist's 5 mM becomes the clinic's 90 mg/pct.

2 Biochemical Molecules Are Generally Fairly Stable

It is desirable that the molecules that are used by a cell for its normal survival and for its special physiological functions should be "somewhat stable." If organic compounds are too unstable, they would not last long enough for a cell to maintain its existence. Fortunately, if these compounds are very stable, our enzymes can still alter them, as this must happen constantly in metabolic interconversions.

Figure 3.4 presents a comparison for 11 different chemical reactions of how fast the reaction occurs by itself, without the help of any catalyst, defined as k_{non}. These values are at the lower part of the diagram, with names of enzymes that perform the same chemical step more quickly.

How did Richard Wolfenden and his colleagues measure these uncatalyzed reactions, if they are so slow? No scientist can wait thousands or millions of years to measure the changes in such hydrolytic reactions. The solution was to perform these experiments at several much higher temperatures, in special vessels that would not burst, as the water in such samples was heated above the boiling point. Such chemical rates are faster at higher temperatures and therefore lower as the temperature is lessened. They then extrapolated the observed k_{non} values measured at these high temperatures to room temperature to obtain the values in Fig. 3.4.

Rate values for enzyme activity, written as k_{cat}, are at the upper part of the diagram. Because these reactions are so much faster, it is routine to make such measurements. Please note that the y-axis (the left scale) has logarithmic divisions for the units, which are in seconds. Choosing the unit of time for such a graph presents an interesting challenge. The age of the Earth is routinely

3 Our Superbees: How Fast Are Enzymes?

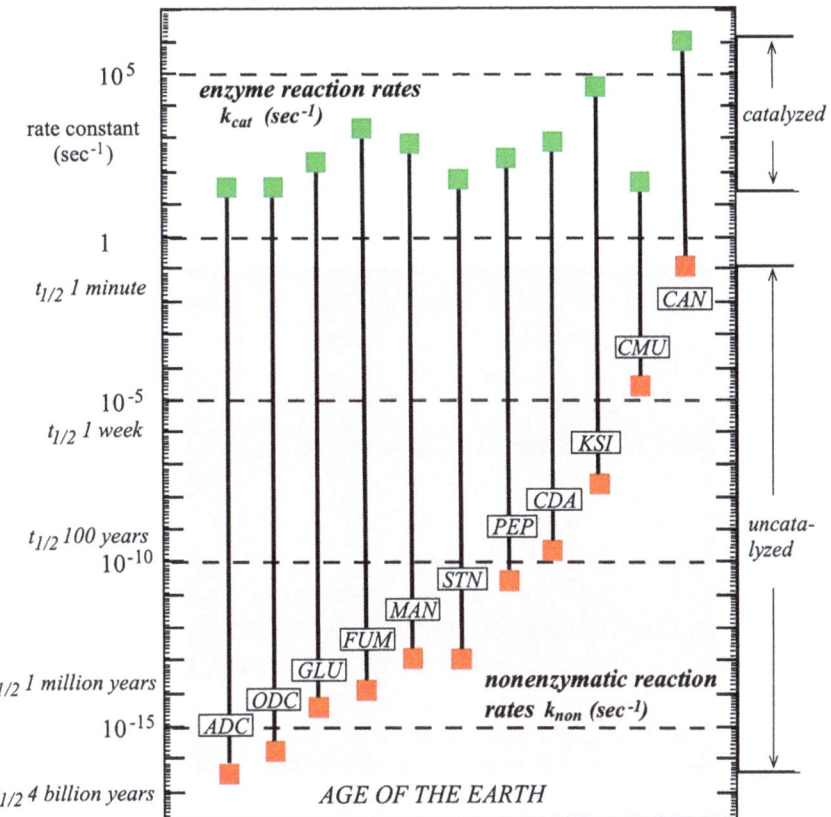

Fig. 3.4 Comparison of uncatalyzed chemical rates, k_{non}, for certain reactions with the rates for enzymes that perform these reactions, k_{cat}. The three-letter abbreviations in the boxes define the enzyme: for example, CAN (at far right) is for *carbonic anhydrase*; STN is for *Staphylococcus nuclease*. (Figure courtesy of Richard Wolfenden)

measured in years, but with this unit k_{cat} values would be astronomical. Carbonic anhydrase (CAN), the enzyme at the top right, has a k_{cat} of 10^6 s^{-1}. If the time units were scaled in years, this k_{cat} value would have to be multiplied by the number of seconds in one year. But with the time scale using units of seconds, the slowest k_{non}, at the bottom left for adenine decarboxylation (ADC), is greater than 10^{-16} s^{-1}. This means that without any enzyme to perform the reaction, the amino acid arginine would require about 3×10^{16} seconds, or about 1 billion years, before it became decarboxylated.

In Fig. 3.2, the time scale is also logarithmic but changes by intervals of 10×; in Fig. 3.4, the horizontal dashed lines change by values of 10^5, or 100,000×, as we look up or down on the graph. Richard Wolfenden, the author of this figure, has helped to make this easier to interpret by adding normal units of time on the left side of this diagram. The lowest major hash mark on the y-axis is at 4 billion years, which is almost as long as the age of our Earth. The right axis shows that the upper values are for catalyzed chemical reactions, while the lower ones are uncatalyzed.

To help readers appreciate the vast span of time depicted, note that the fastest uncatalyzed reaction occurs in a few seconds, at the top right of the diagram, and the slowest reactions, at the bottom, require billions of years. How does this apply to our normal life? We want foods that we gather, or buy, to last long enough for the time it takes from farm to market, and then to our kitchen. But for them to be useful for maintaining our biochemical needs, they must also be easily convertible to some other forms to be used for energy or for biosynthetic reactions.

The stability of foods is also dependent on how they are stored. A bag of dry crystalline sugar is very stable and might last in a pantry, if not consumed, for thousands of years or more. But if the sugar is dissolved in water to make a syrup, it will not last as long. Water itself is a very weak acid and will react with organic compounds and steadily, if slowly, break bonds holding such molecules together. This second process is what is displayed in Fig. 3.4 as uncatalyzed (k_{non}), because the compounds that are substrates for the enzymes indicated will hydrolyze at the speeds measured.

One may see several interesting findings in the experiments of Fig. 3.4. There is a wide range for the stability of different organic compounds. Carbonic acid, cleaved by carbonic anhydrase (CAN), lasts about 5 s. The amino acid arginine, cleaved by arginine decarboxylase (ADC), lasts about 2 billion years. By comparison, the enzymatic rates, shown at the top of Fig. 3.4, do not vary as widely. For this small sample, the enzymes all have a rate, k_{cat}, faster than 100 reactions per second, and the fastest, carbonic anhydrase, has a k_{cat} of about one million reactions per second—for a single enzyme molecule.

In some cases, stability is good. Consider the reaction in the center of this diagram, catalyzed by STN, which represents *Staphylococcus nuclease*. *Staphylococcus* is a common bacterium that is found on the skin of most people. This particular enzyme, a *nuclease*, degrades or breaks down nucleic acids such as DNA. We can then see that DNA, the substrate for *Staphylococcus nuclease*, is remarkably stable, with a $t_{1/2}$ of almost one million years. The designation $t_{1/2}$ is a conventional term for chemical stability and signifies the time "t" for half of the molecules being studied to become degraded. It also means that half of the molecules continue to exist.

Therefore, in a test tube containing your DNA, half of this sample would still be intact after almost one million years. The enzyme, however, can cleave about 100 DNA molecules every second.

3   Evolution Optimizes Biochemistry

The above fact gives us an opportunity to appreciate how evolution selects the best solution. Because DNA is a remarkably stable, long-lasting molecule, it is an excellent choice for storing genetic information, because such a molecule must not only last for the lifetime of a cell, but with germ cells (sperm and ova), the DNA must continue to be stable in the offspring that are produced.

In Table 2.4, we saw that some organisms, such as viruses, use RNA for their genetic carrier. RNA is far less stable than DNA and would be an impossible choice for long-lived organisms. However, it is highly likely that the earliest cells to exist at the beginning of life on Earth actually had RNA for their genetic storage, because the RNA molecule is easier to make. At some time thereafter, perhaps many millions of years, cells discovered how to make DNA.

In this case, "discover" means that several novel enzymes were now being made by these cells, by duplicating a few existing genes and then having their action slightly altered by mutation, so as to make modified nucleotides, meaning that they contain a ribose, the 5-membered ring structure (Fig. 3.5), to which are attached a base at carbon 1' and one or more phosphates at carbon 5'. To help clarify the nomenclature used for nucleotides and their precursors, Fig. 3.6 shows these in more detail.

One might ask how life forms continued to exist with the less stable RNA. After about 4 billion years of life on Earth, viruses are the main organisms that continue to use RNA. And for viruses, this use of RNA enables them to have a smaller size (see Table 2.3). Being small provides a unique and needed benefit: it enables the virus to invade host cells that will help to

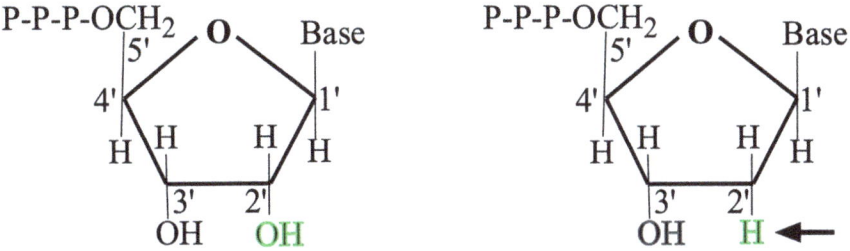

Fig. 3.5   Ribo- and deoxyribo-nucleotides. The arrow points to the side chain of carbon 2' of the deoxyribose, which lacks the oxygen, shown on the ribose at left. Each corner of the 5-sided ring would have a carbon atom (1' to 4'). Any of the 4 bases can be attached at carbon 1' to make the 5 different nucleotides found in nucleic acids

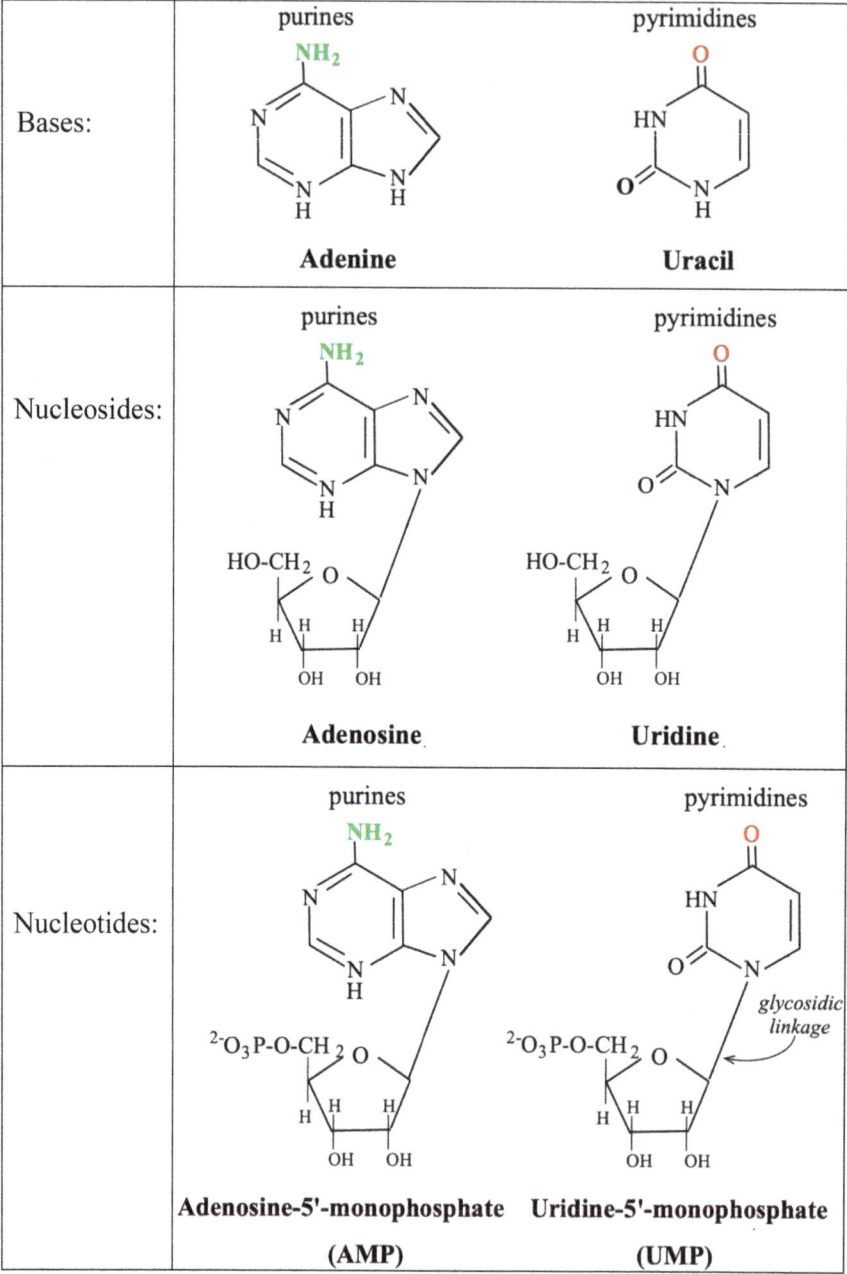

Fig. 3.6 Molecular components of a nucleotide

reproduce many more viruses. The DNA in our cells always exists as a double helical twin molecule because of base-pairing between two DNA strands. RNA is commonly a single-stranded molecule and so requires only half as much space for packaging into a viral capsid. Figure 12.6 shows Covid viruses infecting a mammalian cell, but compared to the size of that cell, the cartoon viruses in this figure are at least 100 times larger than they should be.

While RNA-containing viruses have a genome that is not as stable, they overcome this by the simple ability to reproduce hundreds of copies after each infection. Here, it is good to recall the meaning of $t_{1/2}$. Should even one half or more of the newly synthesized viruses have sufficiently damaged RNA genomes to make those viruses unable to continue, there are so many other successful copies being made that ensure the virus will easily continue to spread.

Resources

A discussion of enzyme rates: https://book.bionumbers.org/how-many-reactions-do-enzymes-carry-out-each-second/

Using exponential notation: https://www.ibm.com/docs/en/zvm/7.2?topic=arithmetic-exponential-notation

4

How Did We Get Here? Starting Conditions for Life on Earth

Abstract Discussion of conditions for life to be possible, the importance of water, and then the appearance of oxygen about 2 billion years ago. The earliest amino acids and nucleotides were formed abiotically (without enzymes) and enabled the earliest bacterial cells to form, with the necessary building blocks to synthesize proteins and RNA. How a nonlethal mutation can become widespread is discussed.

Keywords Water · Oxygen · Abiotic synthesis · Amino acids · Nucleotides · Wide-spread mutations

Before we can appreciate how cells function, we should first consider the initial requirements for life to be possible. Perhaps because our solar system contains four large planets that have extensive gas atmospheres, some science fiction writers have imagined intelligent life forms, mostly composed of gas, living in such environments. This is almost certainly impossible.

The only experience that we actually have with living systems is here on Earth, but we can also be guided by our knowledge of chemistry. While some physicists have speculated about the existence of alternate universes or multiple universes, we have no real evidence for these. It is almost certain that the general principles of chemistry should be the same within this universe.

For life to exist, the following are essential: energy, water, and minerals. While the energy for all life comes originally from our sun, water and minerals emerged billions of years after the universe had formed. Water is very abundant in the universe because it is formed by two of the smallest elements, which came into existence very early. Oxygen and hydrogen atoms form a very stable molecule, and more complex assemblies (Fig. 4.1).

Because of this, the early universe soon had large masses of frozen water. Some of these are still seen today as comets, where the visible tail of these objects is the formation of vapor from the frozen ice as the comet nears the sun.

Such comets would also have impacted the Earth shortly after it formed, and this continued bombardment led to the formation of the oceans that cover almost 70% of our planet's surface. For water to remain liquid, the Earth cannot be too close to the sun (e.g. Venus), where water would boil. And it cannot be too far from the sun (e.g. Mars), where water will freeze. That is why the region in space at the astronomical distance of the Earth from our sun is referred to as the "Goldilocks Zone."

There is one additional feature necessary for a planet to have water: it must be large enough to have the gravity needed to prevent its water supplies from evaporating and diffusing as gas into space. Of the rocky planets, only Earth and Venus are large enough to have maintained any water. Mercury is the smallest rocky planet and also too close to the sun. NASA is still attempting to find evidence of water on Mars, but Mars is about half the size of our Earth

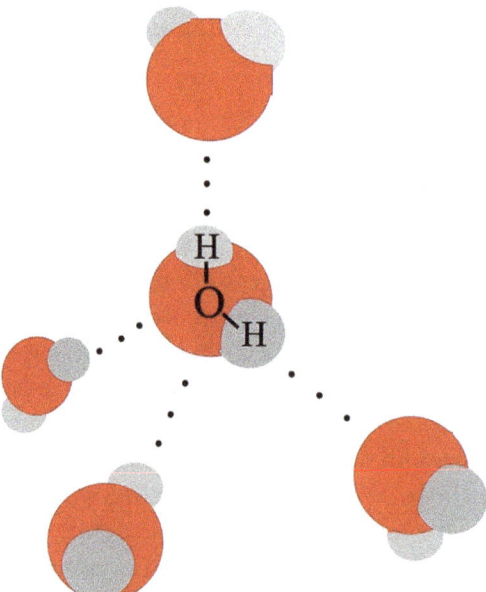

Fig. 4.1 Structure of a water molecule (center). Each water molecule makes hydrogen bonds to four other water molecules

and appears to have lost the water that it may have had in its early years. It has a little ice at the poles.

Because water, in the form of ice, was and most likely still is abundant in the universe, it is almost certain that all the planets and moons would have been struck numerous times with massive ice objects, so as to have the possibility of water. But even our moon, in the Goldilocks Zone for temperature, is not big enough to have retained any of this ice acquired at the beginning of its existence.

Having a moon is also very important. Our moon resulted more than 4 billion years ago when a massive planetary body, about the size of Mars, slammed into the Earth. A large part of this impact object broke up and reformed into a solid moon, which is about one-fourth the size of the Earth. The moon is important to ongoing life for two reasons. The initial impact tilted the Earth off its rotational axis. To understand this, think of the sun with the Earth's orbit around it as a disk or flat plane. Because the solar system formed from a swirling planar dust cloud, the Earth originally would have rotated with its axis perpendicular to this plane. The impact that formed the moon pushed the Earth askew, so that since that impact it has rotated on an axis that tilts by 23°.

This tilt is significant because it changes the part of our planet's surface that is most exposed to the sun as the planet orbits the sun during a year. And this variation in temperature produces the seasons. The other important result of having a moon is that it is close enough and massive enough to have a gravitational pull on our oceans. And as the Earth rotates, this pull changes in how it affects the oceans, resulting in tides.

Tides are very important for producing intertidal zones at the coasts, and this would have enabled the earliest forms of sea life to inhabit these areas. By being intermittently without surrounding water, some of these organisms could have adapted to breathing air directly, and thereby become able to live on land.

1 The Importance of the Early Elements

Astrophysicists have established that the universe is almost 14 billion years old, and that our solar system is about 4.6 billion years old. The universe was almost 9 billion years old before our sun, and its eight planets, came into existence. This may seem like an irrelevant numerical detail, but this time frame has some importance.

At the very beginning of our universe, frequently designated as the "Big Bang," hydrogen was the only element that existed. Hydrogen atoms (H) could collide with each other and become hydrogen molecules (H_2), and these could then aggregate by modest gravitational attraction until they formed massive clouds that condensed and became stars.

The very earliest stars that formed after the beginning of the universe contained almost only hydrogen. Over time, the immense heat of a burning star combined with the immense gravitational pressure led to the chemical fusion of hydrogen atoms to form helium, and with more time, larger elements slowly became more abundant through such fusion processes. The important point here is that stars formed in the first few billion years had none of the much larger elements, which are so important for forming the molecules in all living organisms.

When such early stars died after a few hundred million years by becoming a nova and exploding, all the atomic elements still remaining would be scattered in gas clouds, and such clouds could then recondense and form later-generation stars. These newer stars then started with some elements already being larger than hydrogen and helium, and continued by fusion to make even larger elements, among which metals are very important. When our star, Sol, began to form, the universe was about 9 billion years old. By then, all the early stars had exploded and so scattered their greater abundance of heavier elements. In our own solar system, the inner rocky planets could only form with adequate quantities of silicon and metals. And only then could life become possible. It is highly likely that no life existed until our universe was at least 5 billion years old.

A time frame for these early years in the formation of our solar system and our planet is shown in Fig. 4.2. It is important to see the correct time frame. 4.6 Bya for our planet means that the universe was already over 9 billion years old when our Sun and planets formed.

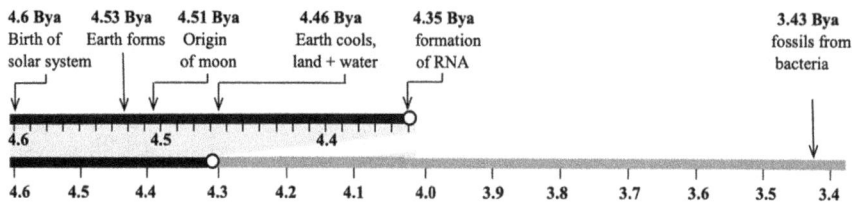

Fig. 4.2 Timeline for formation of solar system and Earth. *Bya* billion years ago

4 How Did We Get Here? Starting Conditions for Life on Earth

Table 4.1 Necessary elements

Atomic number	Element	Symbol	%[a] mass
1	Hydrogen	H	9.5
6	Carbon	C	18.5
7	Nitrogen	N	2.6
8	Oxygen	O	65.0
11	Sodium	Na	0.2
12	Magnesium	Mg	0.1
14	Silicon	Si	
15	Phosphorous	P	0.6
16	Sulfur	S	0.3
17	Chlorine	Cl	0.2
19	Potassium	K	0.2
20	Calcium	Ca	1.3
25	Manganese	Mn	<0.1
26	Iron	Fe	<0.1
27	Cobalt	Co	<0.1
28	Nickel	Ni	<0.1
29	Copper	Cu	<0.1
30	Zinc	Zn	<0.1

[a] % of human body

Of the three main components necessary for life on the previous page, "minerals" is a general term for any source of the 18 elements that are so widely observed on Earth, of which 17 are found in molecules (Table 4.1), and it is certain that life would not exist without them. Six of the eight most abundant elements in our bodies are among the nine most common elements in seawater, where our evolution began.

Hydrogen and oxygen are unusually abundant in human bodies because they form water, and water makes up almost 70% of the volume of each cell. We also have blood, lymphatic fluid, spinal fluid, etc.

A few of the higher atomic elements (not in this table) are infrequently found in some molecules. It is not clear if they are truly essential or represent a fortuitous use of such an element once it became available.

There are many examples of the importance of metals. Most people know that iron deficiency causes anemia. This happens when red blood cells do not have enough iron atoms to bind to their hemoglobin molecules and help them bind and transport oxygen. Magnesium is necessary for kinase enzymes to bind ATP. Calcium is needed to activate the blood clotting cascade. More than 40% of known enzymes bind a metal atom to assist in their function.

2 The Importance of Oxygen

Because oxygen is essential for all mammals, we normally take its existence for granted. But there was almost no oxygen on the Earth after it was formed more than 4 billion years ago. Therefore, the only life forms possible were anaerobic (from Greek: life without air), such as the archaebacteria, the first cells that came into existence. It was about 3.7 billion years ago (3.7 Ga) that a very fortuitous event occurred for all later life—the origin of cyanobacteria, also named blue-green algae.

They had evolved chlorophyll and specific proteins to split water molecules, forming *oxygen* as a by-product. In most photosynthetic organisms, chlorophyll is the key molecule that captures photons and uses their energy to chemically split H_2O molecules, producing ATP (Fig. 1.3) and NADPH (Fig. 4.3).

The oceans are large, and it required two billion years before concentrations of oxygen became adequate to support the first multicellular eukaryotes (from the Greek: *eu* = good; + *karyon* = kernel, i.e. the nucleus), such as sponges. While some bacteria are anaerobic (living without oxygen), all eukaryotes require oxygen. Some of the oxygen in seawater would also diffuse into the atmosphere and support terrestrial life.

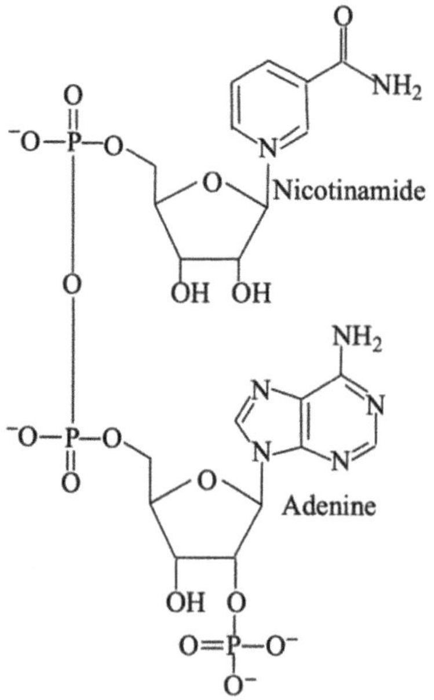

Fig. 4.3 Structure of NADPH (nicotinamide adenine dinucleotide phosphate)

4 How Did We Get Here? Starting Conditions for Life on Earth

It has been estimated that atmospheric oxygen levels need to be about 14–15% for warm-blooded animals to live. We see in Table 4.2 that currently we have an atmospheric oxygen concentration of about 21%. It has also been demonstrated that if oxygen levels were near 30%, this would make the combustion of flammable woods and grasses more probable. Therefore, the present atmosphere is in a Goldilocks Zone with regard to oxygen levels.

Also evident in Table 4.2 is that there were no complex life forms on land until the Earth was almost 2 billion years old. And it was only at about 320 Ma that the first mammals appeared: small rodents like mice. Oxygen provides one additional important function for life on Earth by forming ozone (O_3). This gas occurs in the stratosphere, at very low concentrations of about 5 parts per million of air molecules, which are 78% nitrogen and 21% oxygen, plus traces of carbon dioxide. What makes ozone so important is its ability to absorb UV light. Light in the ultraviolet frequency range is very energetic and can cause damage to almost any molecule it impacts. This makes it possible for it to pass through the outer skin layers and damage DNA in cells existing there. This often causes skin cancer in humans (Fig. 4.4).

The lack of oxygen, and therefore of ozone, in the first 3 billion years of Earth's existence may also have helped with the appearance of novel life forms. Without ozone, the sun's UV rays would have been more plentiful, and therefore the earliest life forms, perhaps in shallow lagoons, would have experienced higher mutation rates, possibly producing novel types of organisms. By the time that there was enough oxygen in the atmosphere to support mammalian life on land, there would also have been enough ozone to protect these animals from excessive UV radiation.

Table 4.2 Appearance of oxygen and O_2-dependent life forms

Time	[O_2] in atmosphere %	[O_2] in seawater	Life forms
> 3.7 Ga	0.0002	3 nM	Archaebacteria
< 3.7 Ga			Cyanobacteria
~3.1 Ga			Origin of aerobic bacteria
~1.7 Ga	2–4	30–60 µM	First multicellular eukaryotes
~320 Ma	>12	> 180 µM	First mammals
Current	21	300 µM	

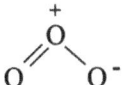

Fig. 4.4 Ozone

3   Abiotic Precursor Building Blocks

We have already seen how nucleic acids are formed as polymers of simpler building blocks, the nucleotides, and that nucleotides have five types of different bases, of the type shown in Fig. 3.5. How did cells know how to make only these distinct types of bases, when dozens are chemically possible? And then choose only those bases now used, out of all the possible chemical bases?

The structure of some bases is sufficiently simple that the atoms within them can actually arrange themselves in the proper orientation and then make bonds between these atoms to produce the final organic structure.

For amino acids, the building blocks for making proteins, there are hundreds of possible types that have been synthesized by chemists. Only 20 are normally used in all natural proteins.[1] The amino acids used in proteins are α-amino acids (Fig. 4.5), where the alpha-carbon (α) is the first carbon after the initial carbon, at right (C). Beta-(β) amino acids are never found in natural proteins.

For the full set of 22 α-amino acids, the side chain attached at "*R*" (for residue) can be any atom or complex of atoms. In the protein-forming amino acids, there is only one more carbon in the backbone—the linear sequence of atoms—and this binds two oxygen atoms as shown in the dotted box (Fig. 4.5). This is called a carboxylic group. It is this carboxylic group that makes these compounds acids.

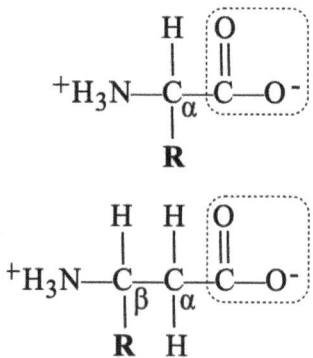

Fig. 4.5 Structures of an alpha (α)-amino acid (top) and a beta (β)-amino acid (bottom). *R* is the side chain, which is different in each amino acid

---

[1] Two extra amino acids, selenocysteine and pyrrolysine, are found in some proteins.

4 How Did We Get Here? Starting Conditions for Life on Earth

Amino acids can contain more carbons in the backbone, "N—C_γ—C_β—C_α—C—O". In such a structure, the carbons are normally designated by the Greek alphabet. When the side chain, R, is at carbon β, these would then be beta-amino acids, and so on. There are hundreds of possible amino acids.

The answer to the question above is that the simplest amino acids can form spontaneously in any aqueous pool that contains the elements necessary for their formation in sufficient abundance, so that such fortunate assemblies were statistically likely to happen. And this is consistent with the interpretation that life was unlikely to arise until at least 5 billion years after the Big Bang.

This means that the earliest and simplest forms of bacteria had some of these precursor building blocks already available for use in making the first RNA molecules and some original small proteins. Because this type of synthesis occurs without living cells, it is referred to as *abiotic*. The vitamins that many of us take for their health benefit do this by often acting as cofactors for specific enzymes in their chemical reactions. Many such vitamins also originally occurred abiotically and therefore were available when living bacterial cells originated. Currently, many types of organisms have evolved enzymatic pathways to synthesize such vitamins themselves, making them less dependent on their diet.

As an interesting example, we know that vitamin C (ascorbate) is readily obtained from citrus fruits because they contain vitamin C in abundance. Our cats and dogs do not need to drink orange juice frequently because they do not need an external source for vitamin C; they can synthesize it themselves. There are several different chemical steps in the synthesis of ascorbate, each requiring a different enzyme that is coded by a different gene. In humans, very early in their evolution, one of these genes became altered and no longer produced the needed enzyme. But early humans lived in Africa, where fruits with vitamin C were abundant. Therefore, this genetic defect caused no problems until humans migrated to northern Europe, where fresh fruit was not as readily available.

When we speak of the human species having experienced a genetic mutation, this may be interpreted as if, in some mysterious fashion, all humans somehow experienced the identical error in that gene in a short period of time. Of course, this is impossible. The reason that a mutation became established in our entire species is that a few hundred thousand years ago, there were only a limited number of people living in Africa. This number may have been less than 5000 people. If just one tribal chief had such a mutation, then because he was a dominant male, possibly having many more wives than most men in his tribe, he could have had many more offspring inheriting this mutation, and they in turn would pass it along to their offspring. If these offspring were otherwise healthy, and the mutation was not lethal or even deleterious, then within four or five generations it might become dominant in this tribe.

The possibility of such large-scale polygamy is supported by the fact that even in the early twentieth century, there were native kings in Africa who had more than a hundred wives. The Hebrew Bible also relates how King Solomon had more than a hundred wives 3000 years ago.

Resources

Formation of the solar system and Earth: https://news.uchicago.edu/explainer/formation-earth-and-moon-explained

Formation of the atmosphere and oceans: https://www.britannica.com/science/geologic-history-of-Earth/Development-of-the-atmosphere-and-oceans

Development and changes in the Earth's atmosphere:
https://forces.si.edu/atmosphere/02_02_00.html

Abiotic synthesis of early molecules:
https://bio.libretexts.org/Bookshelves/Biochemistry/Fundamentals_of_Biochemistry_(Jakubowski_and_Flatt)/Unit_IV_-_Special_Topics/30%3A_Abiotic_Origins_of_Life#:~:text = Abiotic%20Synthesis%20of%20Amino%20Acids,in%20meteors%20suggesting%20the%20possibility.

5

The Nuts and Bolts of Metabolism: How Cells Function

Abstract Description of sources for metabolic energy, which mainly lead to the synthesis of ATP, the most important energy molecule within cells. The abiotic formation of pyrrole enabled the biotic formation of porphyrins, which include chlorophylls in plants and hemes in animals. The importance of thermodynamics and chemical equilibria is described. Energy barriers between different compounds provide a limit to how easily or quickly they can be changed into the other one. Explanation of why some amino acids are essential. Description of the likely origin of life and of LUCA, the last universal common ancestor.

Keywords ATP • Porphyrins • Chlorophyll • Heme • Proteins • Polysaccharides • Lipids • Nucleic acids • Equilibrium • Thermodynamics • Free energy • Essential amino acids • Last universal common ancestor

1 Overview of Metabolism

As briefly discussed in Chap. 1, living organisms are composed of lifeless molecules, and when these molecules are isolated and examined individually, they conform to the same physical and chemical laws that describe the behavior of inanimate matter. Yet living organisms possess extraordinary attributes clearly not exhibited by any random collection of molecules. Three things distinguish living organisms from inanimate matter.

First is their degree of chemical complexity and organization—thousands of different molecules make up the cell's intricate internal structure and enable it to function.

Second is the ability to extract, transform, and use energy from their environment—this enables organisms to build and maintain their intricate structures and to do mechanical, chemical, and osmotic work. This dynamic state must be maintained out of equilibrium, and energy input is required to do this. Inanimate matter tends to decay toward a relentlessly more disordered state, to come to equilibrium with its surroundings.

Finally, living organisms have the capacity for precise self-replication and self-assembly, properties that are the very essence of the living state. Each of the thousands of components of an organism has a specific function—this is true not only for macro-sized structures or subcellular organelles, but for individual chemical compounds as well. The interplay among chemical components of a living organism is dynamic; changes in one component cause coordinating or compensating changes in others, with the whole ensemble displaying a character well beyond that of its individual constituents. The entire collection of molecules carries out a complex program, the end result of which is reproduction of that program, as well as self-perpetuation of that collection of molecules, by producing offspring; in short, life itself.

2 A Brief Look at Energy Formation

The original source of energy for all life forms on Earth is our sun. As briefly mentioned earlier, it was the appearance in cyanobacteria of novel complexes, formed by proteins plus chlorophyll, that enabled them to capture the energy in photons and use that energy to synthesize simple sugar molecules, which in turn are used by the cell to make the common energy molecule ATP (Fig. 5.1).

One should appreciate how remarkable this process is. Water is immensely abundant on our planet. Carbon dioxide (CO_2) is the major waste product from the normal oxidation of fuels for energy. Photosynthesis uses these two very available molecules and converts them into sugars, a metabolic energy source, and also oxygen, which is essential for more complex life. The key to this process is chlorophyll, a porphyrin compound (Fig. 5.2) that appeared very early in the evolution of life.

The four corners of a porphyrin molecule are the pyrrole rings, which are sufficiently simple that they form without the benefit of living organisms. This abiotic availability made it possible for early cells to use pyrroles by joining them into a more elaborate structure known as a porphyrin (Fig. 5.2).

5 The Nuts and Bolts of Metabolism: How Cells Function

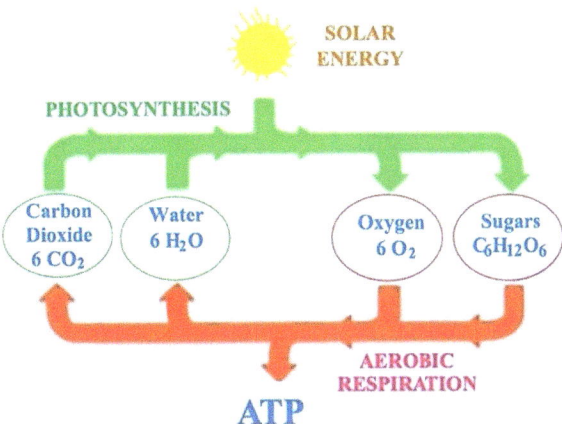

Fig. 5.1 All energy for life comes from the sun

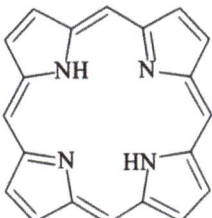

Fig. 5.2 Structure of porphyrin

Porphyrins, in turn, are the building blocks for making many functionally important molecules, among which are chlorophyll in plants, heme in animals (Fig. 5.3), and vitamin B_{12}. We see in Fig. 5.3 that it takes minor modification of a porphyrin to make it bind with a different metal to assist in its specific function. A few examples: Chlorophyll binds magnesium (Mg^{II}) and uses this metal atom to capture the energy in photons, which is then transmitted to other members of the photosynthetic complex to split H_2O and make NADPH (Fig. 4.2), which in turn is a cofactor in the synthesis of sugars. Hemes, attached to hemoglobin, contain iron atoms (Fe^{II}) to assist in binding oxygen in the lungs, which is then released in the capillaries to diffuse into adjacent tissues. Vitamin B_{12} (not shown) is also known as cobalamin because it is an amine compound that contains the metal cobalt. As a vitamin, it acts as a cofactor with numerous enzymes in synthesizing myelin on neurons, developing red blood cells, etc.

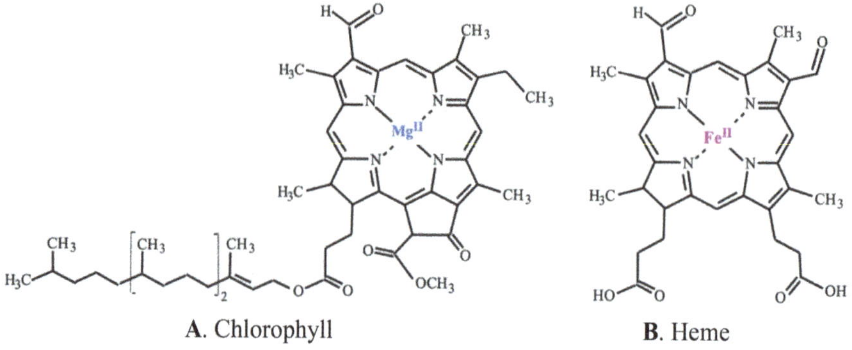

A. Chlorophyll   B. Heme

Fig. 5.3 (a) Chlorophyll (b) Heme

An interesting point here is that evolution is an opportunistic process. Once a useful molecule or structure has been created, it is then frequently used for multiple similar but different functions.

3   Metabolism Recycles Molecular Compounds

The ready availability of an external energy source, our sun, then made it possible to acquire energy continuously by the formation of simple sugar molecules such as glucose. All life forms are able to use glucose, and other sugars, to form the high-energy molecule ATP, and with this metabolic energy source, all possible metabolic compounds can then be synthesized, as shown with the simplified scheme in Fig. 5.4.

Similar molecules are used by the cells as energy precursors and also as cellular components; this may be seen as a form of metabolic recycling. Due to their somewhat different chemical properties, we have these major types of molecules in our cells:

Proteins:	Polymers of amino acids
Polysaccharides:	Polymers of simple sugars
Lipids (fats):	Polymers of ethane and acetate
Nucleic acids:	Polymers of nucleotides

As diagrammed in Fig. 5.4, metabolism has two opposing processes. *Anabolism* refers to the formation or synthesis of molecules, while *catabolism* is the opposite—the breakdown or decomposition of compounds. Without energy, any cell or chemical system quickly ceases anabolic processes, and

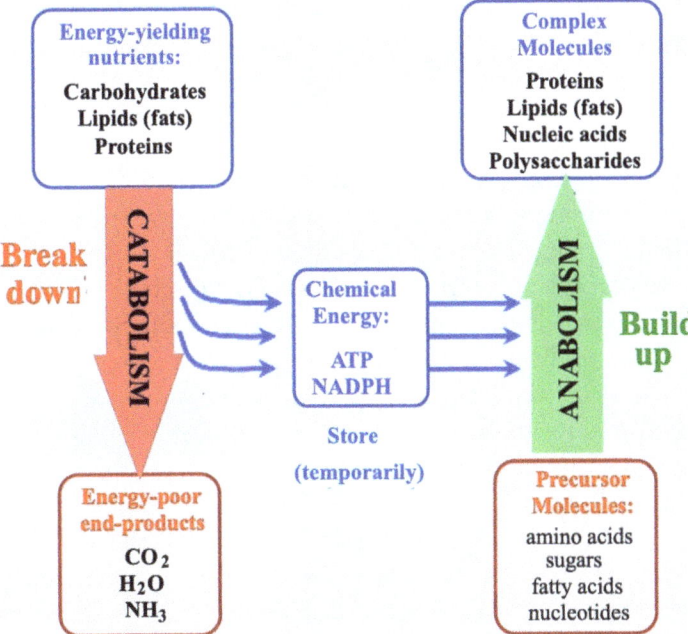

Fig. 5.4 Summary of metabolism

catabolism, usually by the hydrolysis of these compounds with the surrounding water, drives the steady breakdown to the simplest end products.

4 The Importance of Thermodynamics and Equilibria

To really understand and appreciate the benefit of enzyme catalysis, we need to first have a brief consideration of *thermodynamics*, from the Greek *therm* (for heat) and dynamic for causing or having an effect on. This topic was first discussed in the age of steam engines, when heat was first used to boil water and the steam created could move pistons to produce mechanical work. This then also produced the need to understand equilibria, when a system is at rest.

"Rest" may be a misleading term. To many people, it suggests that nothing is happening. But more correctly, it means that there is no net change. As a simple metaphor, consider a store with a revolving door at the entrance. In the middle of the day, it may have many customers inside—in chemical terms that would be the concentration of customers. As 5 PM approaches, customers may begin to exit more frequently in order to get home for dinner, etc. The concentration of customers inside is therefore declining. But some shoppers may have different personal time schedules and so continue to enter the store. At some point, the number of customers leaving will be the same as the number of customers entering, so that the concentration of customers inside now remains constant. That would be equilibrium. But note that there is still a lot of movement out of the store and also into the store, meaning that there is a lot of action, but that it is balanced or equal.

An important aspect of thermodynamics is the concept of energy, especially "free energy," which is the energy in a system available to be used for any work. We have already noted that ATP is the most common high-energy molecule for cellular metabolism. In contrast, water has no energy. The fact that steam can be used for mechanical work does not disprove this last statement because steam is caused by heating water with some external energy source, such as coal or electricity.

Figure 5.5 illustrates how chemists view free energy, designated by the letter "G," to honor Willard Gibbs, who first described equations for explaining this subject. A change in "G" is designated as ΔG (delta G). Shown in Fig. 5.5 is the chemical transition of a compound, A, toward two possible products, B or C. The short horizontal lines represent a constant energy level. The possible product "C" is at a higher energy level, shown visually in the left graph and denoted by "+ΔG." Only a few molecules of A might be able to undergo the energetically uphill reaction, but they will readily go downhill to B. The right panel shows the reverse reaction, where B is converted back to A. Here, B is at a lower energy than A, so only a few molecules of B will convert back to A.

In Chap. 3., we already noted that for molecules to be useful in life, they cannot be too stable or too unstable. The term "stable" really refers to their free energy level, as depicted more exactly in Fig. 5.6, which illustrates the

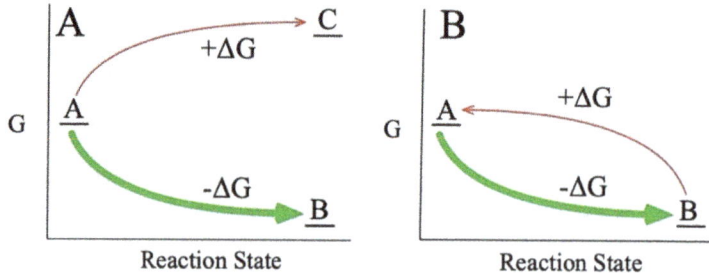

Fig. 5.5 Chemical reactions are only favorable when they go to a lower energy level, depicted by green arrows. G is the symbol for free energy. ΔG is a change in energy

5 The Nuts and Bolts of Metabolism: How Cells Function

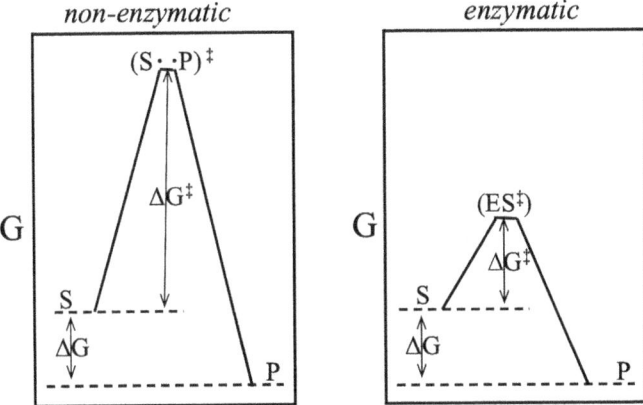

Fig. 5.6 Energy barriers without and with an enzyme. *G* free energy, *S* substrate, *P* product, *ΔG* change in free energy, *ΔG‡* energy barrier for the transition state

energy barrier between S, the substrate for an enzyme, and P, the product. To make this less abstract, let us use the analogy of how water is formed and moves on our planet. Water in the oceans evaporates due to energy from sunlight. Once in the atmosphere, such water vapor may drift over a mountain range, where in winter it will be cold enough to cause them to crystallize and fall down as snow, accumulating as ice on the mountain top or the mountain sides. Here, water will be at a local equilibrium (no change in altitude) until it is melted by summer heat.

This liquid water is in a gravitational field, like the energy field represented in Fig. 5.6, and will only flow downhill. The small streams may accumulate briefly in lakes (a local equilibrium), but with continued melting, the lakes overflow, and the water continues downhill. It will continue to flow until it reaches the sea, which would be the lowest energy level of its gravitational field.

In Fig. 5.6, S is at a higher energy level than P; therefore, it is normal for S to be converted to P, which is energetically downhill. But, as noted earlier, organic compounds need to be somewhat stable, and that is represented by the energy barrier between S and P. Here $(S \cdots P)^{\ddagger}$ represents this intermediate position with the symbol $\ddagger$. This is also called the *transition state*, as the molecule S is chemically converted to P. The higher this energy barrier between S and P, the less likely it is for the change to occur, as if it had to go over a mountain pass.

Now observe how much lower this energy barrier is when S binds first to an enzyme, forming an ES complex. There is still a transition barrier, but now it is energetically much lower, denoted by (ES)‡, meaning that the chemical reaction will proceed more easily and more swiftly.

To summarize: in the absence of an enzyme, energy barriers are normally sufficiently high that the compounds remain almost completely unchanged. The reactions are still possible, as displayed in Fig. 3.4, but they happen far less frequently. But with the appropriate enzyme available, these chemical reactions now proceed at a much faster rate because the enzyme has lowered the transition state energy barrier ($\Delta G^{\ddagger}$), and therefore normal metabolism and living cells are possible.

However, even with the energy barrier, some molecules of P will constantly be converted back to S. In Fig. 5.7, this is diagrammed with a large bold arrow from S to P and a very light arrow from P to S because that is much more unfavorable.

This is because the energy barrier is much higher for P. Conversion of S to P must also still surmount the energy barrier, but it is an energetically more favorable reaction.

Note that with this reaction at or near equilibrium, the concentration of S is now much smaller than P. It is this accumulating concentration of P that enables the back reaction to become equal to the forward reaction. With many more molecules of P, it is statistically possible for a fraction of them to go over the energy barrier.

Returning to our analogy with water evaporating: at the surface of the ocean, there is frequently enough sunlight (= energy) to cause some water molecules to leave the aqueous state and enter the vapor state (i.e. form a gas or mist), even though this is energetically uphill. And because the oceans have so much water, enough of it will recycle to form ice on a mountain top, etc.

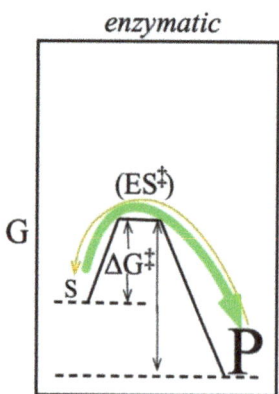

Fig. 5.7 The reverse reaction with *P* converted back to *S*

We should also note that the chemical reaction depicted in Figs. 5.5, 5.6, and 5.7 has been simplified, because most chemical reactions contain two or more substrates and two or more products. Only *isomerases* react with a single substrate and reposition some atoms on that S to make a slightly altered P, but having the same molecular weight.

5 Is there an Energetic Reason for the Common Molecules in Life?

At this stage, it will be more understandable how abiotic molecules came into existence. Their occurrence was possible in part due to the fortuitous availability of the elements needed to produce them, but was also limited by the energy levels possible for such assemblies, i.e. mostly at lower energy levels in terms of their total formation. Examples of this have already been mentioned with the simplest compounds that occurred abiotically. Higher energy levels usually occur with the increased complexity of a compound.

With amino acids, this can be seen with two very different examples: glycine and tyrosine (Fig. 5.8). In these structures, each horizontal or vertical line represents a chemical bond formed by a chemical reaction to join the adjacent atoms. Also, in each structure, the horizontal row of atoms at the top is the same and is the *backbone* of an amino acid.

The 6-membered ring in the structure of tyrosine represents a shortcut by chemists, where each of the six corners has a carbon atom (i.e. "C"), which is omitted in the diagram for simplicity. These side chains for any amino acid are attached at the first carbon, normally the α carbon, and are also called the residue, or **R** in diagrams.

It is evident that the formation of glycine is much simpler, and therefore thermodynamically and statistically more likely to occur in the absence of

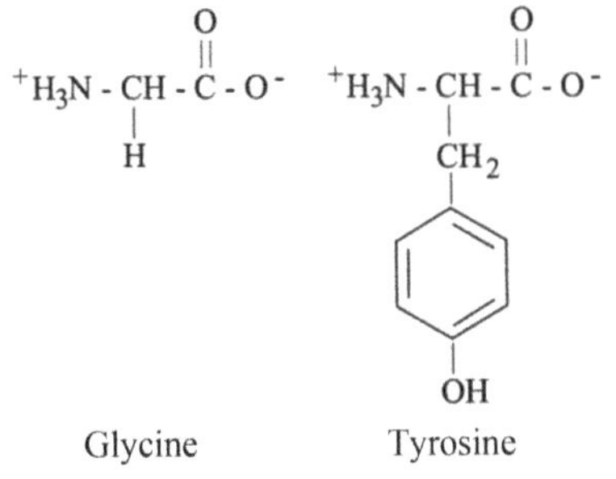

Fig. 5.8 Amino acid structures

enzymes. This helps us to see why some amino acids have become essential. If there are many chemical steps involved in their synthesis, this would require more genes to code for the different enzymes needed, and then it would be more likely that one (or more) of such genes incurs a mutation and one of these enzymes is no longer made.

It must be remembered that not all mutations are lethal or even detrimental. We saw this with the need for vitamin C in humans when they could no longer make ascorbate but still had easy access to fresh fruits and vegetables.

The easy ability to compensate for altered or lost genes is summarized in Table 5.1, which compares nonessential amino acids and essential amino acids by how many enzymatic steps are required in their synthesis. It is evident that the nonessential amino acids are the simplest, requiring the least number of steps for their formation. One must understand that there is a possible misinterpretation if one compares Fig. 5.7 and Table 5.1 for glycine, as an example. Figure 5.7 shows each bond in the structure accurately, and each would require a nonenzymatic chemical reaction. However, in a human cell, glycine is easily formed in a single step by adding an amino group (NH_3) to acetate (CH_2COO^-).

The most complicated amino acid among the nonessential ones is arginine, requiring four steps to synthesize. Even though arginine is just a little bit more complicated, it is probable that its loss was not possible because arginine is so important for the removal of toxic ammonium.

Careful readers will notice that there are only 8 abiotic amino acids used by living cells, meaning that when the earliest bacterial cells came into existence 4 billion years ago on this planet, these were the only ones available to make the first proteins. Therefore, these first proteins had to be somewhat simpler than what is possible today, but presumably a limited number of functional

Table 5.1 Essential and nonessential amino acids in adults

Abiotic AA	Nonessential Amino acids	# steps in synthesis	Essential Amino acids	# steps in synthesis	Abiotic AA
Alanine	Alanine	1	Cysteine	8	
	Arginine	4	Histidine	10	
	Asparagine	2	Isoleucine	6	
Aspartate	Aspartate	1	Leucine	8	
Glutamate	Glutamate	1	Lysine	10	
	Glutamine	3	Methionine	7	
Glycine	Glycine	1	Phenylalanine	10	
Proline	Proline	3	Threonine	5	Threonine
			Tryptophan	11	
Serine	Serine	3	Tyrosine	16	
			Valine	6	Valine

structural proteins and enzymes were possible. Later, by the process of gene duplication followed by some mutations, newer enzymes came into existence that began to synthesize some of the "missing" amino acids.

There continues to be debate about the details of the origin of these earliest cells. It has been mentioned that it is fairly certain that the earliest cells used RNA to make the genes coding for the new, and probably simpler proteins. The time spans for cells to discover DNA and convert to using it for storing genetic information have not been determined. Some readers will be aware that scientists have recovered ancient DNA samples, usually from bones or teeth. The oldest such DNA recovered, that had sufficient integrity to be sequenced in a lab, goes back about one to two million years. In Fig. 3.3, we saw that the $t_{1/2}$ for DNA is about one million years. Scientists continue to find some older fossils, such as fossilized bacterial cell mats, and DNA fragments from these have been helpful in defining the earliest common ancestor for all life forms, referred to as LUCA (Last Universal Common Ancestor).

Figure 5.9 depicts a timeline to make it easier to comprehend how the various processes described above occurred, making it possible for the various life forms to appear. The appearance of LUCA, our universal common ancestor, has been estimated with reasonable certainty at about 4.2 Ga. If the Earth was impacted at about 4.5 Ga by a large wandering planet (i.e. not established in a normal orbit around the sun) that resulted in forming our moon, then within 100 million years or more, our planet cooled sufficiently for water to become abundant, thereby making the formation of abiotic compounds possible at about 4.3 Ga. The various earliest life forms then arose during the next 100 million years or so, and by 4.2 Ga, LUCA became established as the most successful of these.

Figure 5.9 suggests, with red lines and one bold green line, how DNA was exchanged. At the very bottom of the upper right figure, the earliest organisms may not all have changed to using DNA and may also have had a few other features to make them separate and distinct. But one cell line early on became the most efficient and formed the beginning for all other life forms.

The cell line labeled LUCA has this distinction because it had the same genetic code seen in all organisms today. It would have had all the features depicted for bacterial cells, as shown in Fig. 2.1. After LUCA, there is an initial split into two separate evolutionary lines: archaea and bacteria. The tree of life depicts the evolution of the three separate kingdoms of life: archaea, bacteria, and eukaryotes (from the Greek: *eu* = good; + *karyon* = kernel, i.e. the nucleus). The heavy green line connecting bacteria to eukaryotes is a simple way of depicting how eukaryotes first appeared when some species of bacterial cells infected archaeal cells, and by chance did not destroy the cell that they

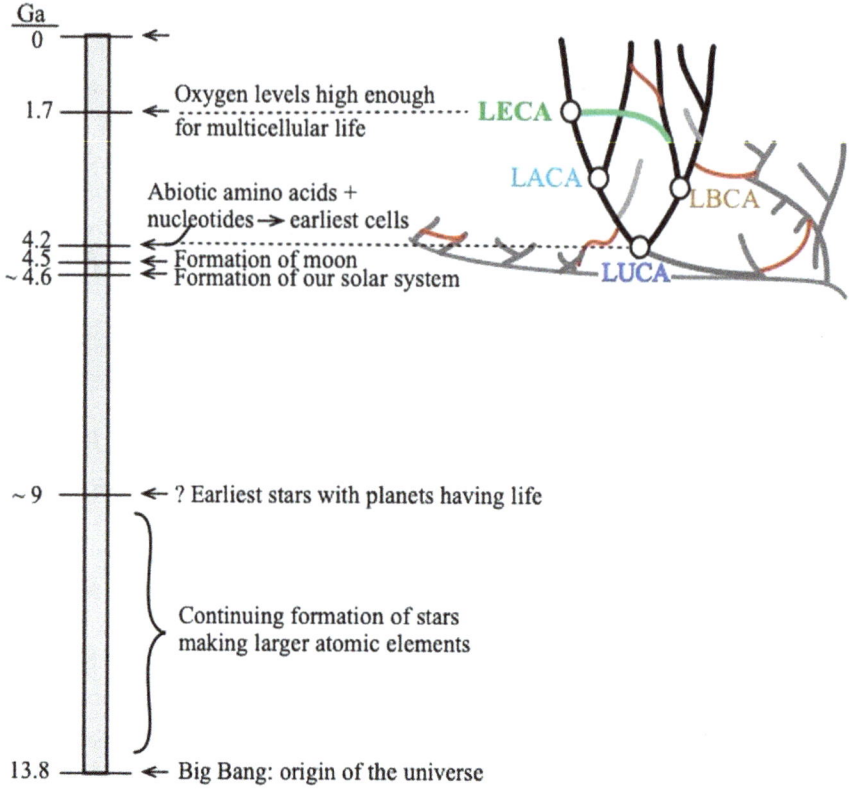

Fig. 5.9 Timeline for the formation of stars and planets, and the appearance of living cell lines. *LACA* Last Archaeal Common Ancestor, *LBCA* Last Bacterial Common Ancestor, *LECA* Last Eukaryote Common Ancestor, *LUCA* Last Universal Common Ancestor. Red lines suggest gene transfer between species. A green line shows symbiosis of bacterial cells and archaeal cells to form eukaryotic cells with mitochondria

had infected, but took up residence inside the larger archaeal cell. With evolutionary modifications, the bacterial cell became converted into the mitochondrion that all eukaryotic cells have. This was a very successful symbiosis, as the mitochondrion evolved to become the major source for making ATP.

6 How Is Metabolism Organized?

Most people are familiar with the term "organic" and think that it applies to how fruits and vegetables are grown. But for chemists, this word has always defined molecules or compounds formed with the element carbon, or carbon chemistry. Carbon was available early, as it is element #6, and it has a very important feature in that it is able to make bonds with four other atoms. This enables remarkable complexity in the formation of organic compounds when various atoms bind with carbon. Amino acids are good examples (Fig. 5.8).

Metabolism may also easily be separated into the formation and degradation of proteins, or lipids, or sugars, or nucleic acids, etc. These separate realms of metabolism may be studied by themselves, but they function in a very coordinated fashion within our cells, while still retaining special functions suitable for their chemical structure.

Resources

Introduction to thermodynamics: https://library.fiveable.me/chemical-kinetics/unit-10/relationship-kinetics-thermodynamics/study-guide/vwkIR5Z5OEj5vR9r

Explanation of enzyme transition states: https://library.fiveable.me/biophysical-chemistry/unit-4/transition-state-theory-biological-systems/study-guide/6fK5Tn1NXciYRFhe

Explanation of abiogenic production of simple molecules and the appearance of the LUCA, the Last Universal Common Ancestor. https://en.wikipedia.org/wiki/Abiogenesis

6

The ABCS of Protein Structure and Synthesis

Abstract Description of protein structures with modules and domains, and protein folding. Use of sequence alignments to identify consensus structures, and sometimes the identity of newly found proteins, and also how related different species are. The importance of hydrophobic amino acids, which energetically prefer to be in the interior of proteins, in determining the folding of a protein chain into a folded structure. Larger proteins have more domains and therefore have more possible functions. The importance of proteins being able to change their conformation. An extended linear protein chain has positions for extra processes. Ubiquitin attracts the proteasome to degrade unfolded proteins or unnecessary proteins. A brief description of covalent and noncovalent bonds that stabilize the final protein structure. Description of protein synthesis using ribosomes and tRNAs.

Keywords Alpha-helix • Beta-strand • Types of collagens • Domains • Modules • Protein conformation • Signal peptides • Proteasome • Ubiquitin • Covalent bonds • Noncovalent bonds • Protein synthesis • Ribosomes • tRNAs

We have two major types of proteins based on general structure. The majority of intracellular proteins are globular, meaning that they fold into a somewhat rounded shape (see Fig. 6.1). A second type are the fibrous proteins, because of their tendency to form such structures. These are very common among the structural proteins:

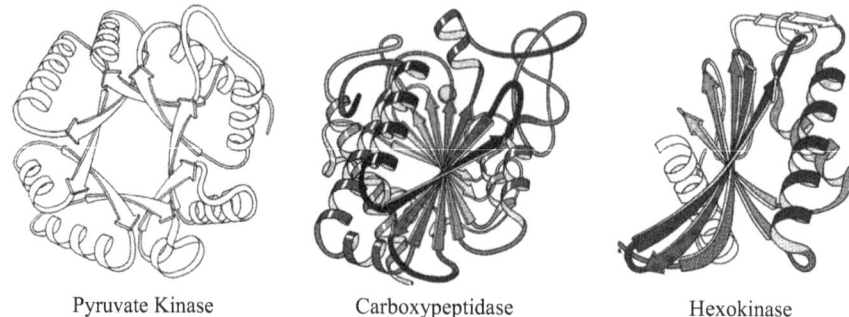

Pyruvate Kinase   Carboxypeptidase   Hexokinase

Fig. 6.1 Schematic examples of fully folded proteins. Note the secondary structure elements, denoted by helices and arrows for β-strands. (Source: Jane Richardson; Advances in Protein Chemistry (1981))

Table 6.1 Types of collagens

Type I	Makes up 90% of the body's collagen. Type I is densely packed; used to provide structure to the skin, bones, tendons, and ligaments
Type II	Found in elastic cartilage, which provides joint support
Type III	Found in muscles, arteries, and organs
Type IV	Found in the layers of the skin
Type V	Found in the cornea of eyes, some layers of skin, hair, and in the placenta

Myosin and fibrinogen are α-helical.

Collagens form a triple helix, which can be flexible, as in skin.

Wool fibers (α-helix) are flexible and extensible, as in sweaters.

Silks are β-pleated sheets; they have unusual strength, great rigidity, but low extensibility.

Structural proteins are very abundant in specific tissues and provide more body mass. As examples, actin and myosin form the contractile filaments that provide muscle action, and we have a lot of muscle, especially in our limbs. Collagen provides the structural protein for forming a mesh or mold that can be filled in to make cartilage in joints and bones in our skeleton. It is so widely used that we have five major types of collagens (Table 6.1).

1 Types of Protein Structure

The use of the flat arrow for beta strands and the spiral coil for alpha helices, shown below for visualizing protein structure, was first introduced in the 1970s by Jane Richardson at Duke University and has helped scientists to be able to understand three-dimensional structures more easily and thereby compare how similar two different proteins might be. There are four types of protein structure:

Primary structure: A linear amino acid sequence, e.g.: GACKLEMD ⋯
Secondary structure: Two major types, shown below and on the next page (Figs. 6.2 and 6.3).

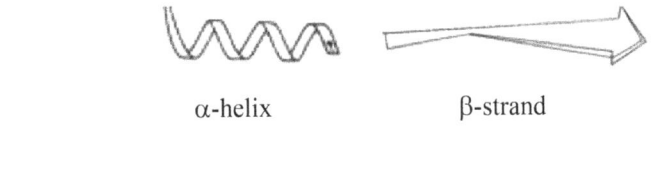

α-helix   β-strand

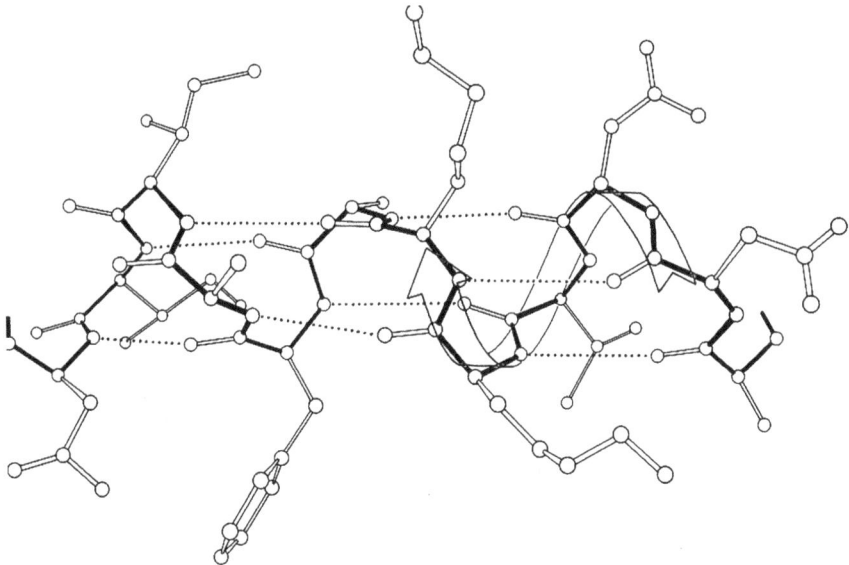

Fig. 6.2 Atomic structure of an α-helix. Each small circle represents an atom. Note the overlaid helical ribbon, similar to the drawing above this figure, which helps visualize the continuation of the spiral or helical structure. This represents the backbone of the protein chain. The extensions above and below are the side chains of amino acids. Dotted lines are shared hydrogen bonds, *within the same helix*, which stabilize this structure. (Source: Jane Richardson; Advances in Protein Chemistry (1981))

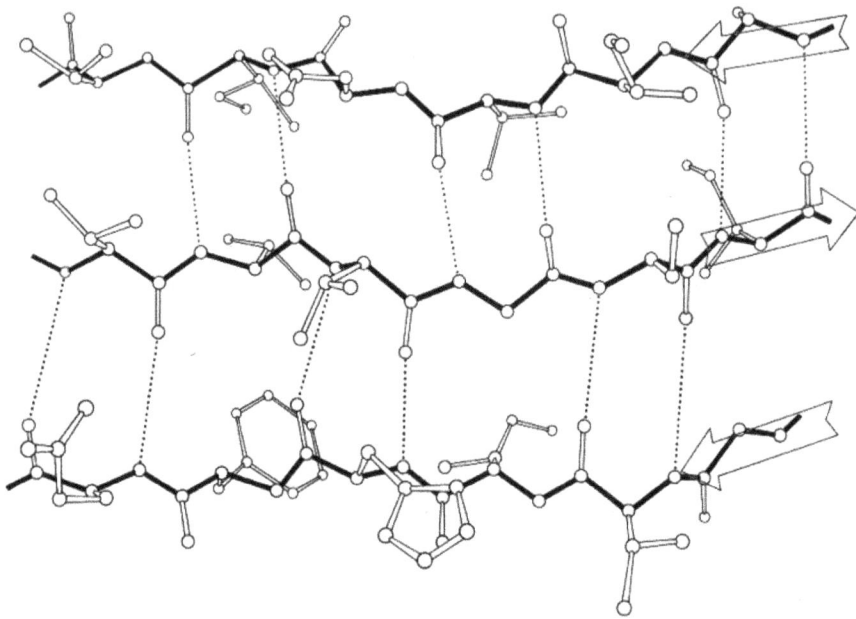

Fig. 6.3 Atomic structure of a β-sheet containing three β-strands. The three schematic arrows suggest the individual strands and their directions. The upper strand will fold back on the left and continue as the middle strand and then fold back on the right and continue as the bottom strand. Dotted lines are shared hydrogen bonds *between strands* that stabilize this structure. (Source: Jane Richardson; Advances in Protein Chemistry (1981))

Tertiary structure: The fully folded protein, as diagrammed (Figs. 6.1 and 6.4). Colors in Fig. 6.4 indicate that the initial linear unfolded protein chain has the properties for folding into two structural domains. Protein domains vary in size (Fig. 6.5). The term domain is normally used for proteins that contain two or more domains. Examples of fully folded tertiary structures are shown in Fig. 6.1.

Sequence alignments, as shown in Fig. 6.6, establish the degree of relatedness among any two or more organisms. This is possible because amino acid changes, due to changes in the DNA codon(s), occur at a fairly slow but standard rate and they occur randomly. The more amino acid differences that occur between two species for the same proteins, the further back in time they became separated.

All important functions of proteins are a consequence of the properties of the amino acids, defined by their individual side chains.

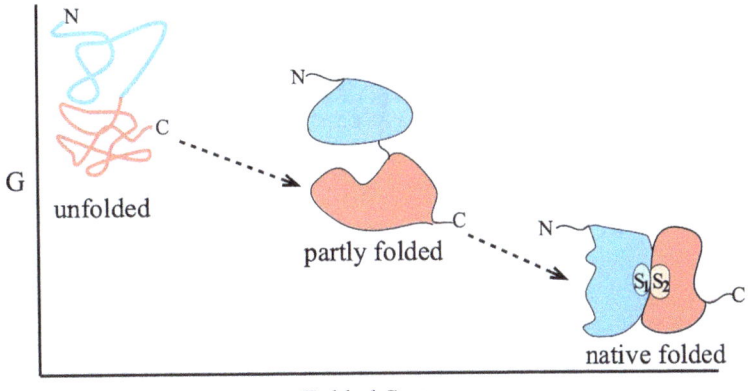

Fig. 6.4 Folding steps from the initially synthesized protein chain, via intermediate states, to the final native folded protein with two domains

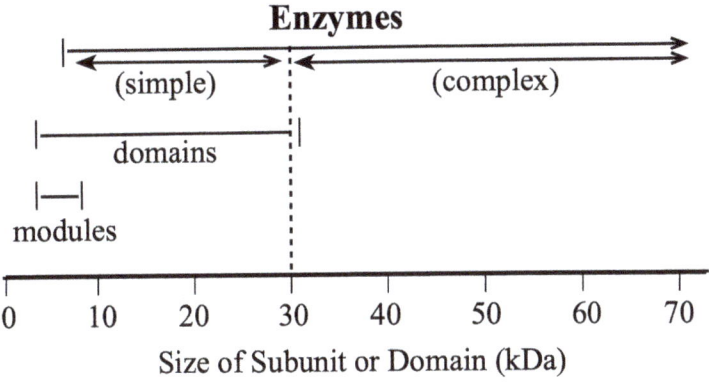

Fig. 6.5 Variation in the size of proteins and in the size of structural components. The demarcation at 30 kDa is an approximation for single-domain enzymes

We see the name of each amino acid, as well as the three-letter abbreviations used, which easily become obvious. The one-letter code is not as obvious, but it is actually more widely used in any analysis dealing with actual amino acid sequences. Using single letters facilitates the process of sequence alignments when multiple sequences are being compared (Fig. 6.6).

Such an absolute time rate may be questioned, but when it is done for many genes/proteins from the species being compared, there is then an average time that can be validated by archeological data. The earliest mammals only arose 300 million years ago, and because it is now also possible to

```
human  KKASKPKKAASKAPTKKPKATPVKKLAATPKKAKKPKTVKAKPVKASKPKKAKPVK
chimp  KKASKPKKAASKAPTKKPKATPVKKLAATPKKAKKPKTVKAKPVKASKPKKAKPVK
mouse  KKAAKPKKAASKAPSKKPKATPVKKPAATPKKAKKPKTVKAKPVKASKPKKAKPVK
rat    KKAAKPKKAASKAPSKKPKATPVKKPAATPKKAKKPKTVKAKPVKASKPKKAKPVK
cow    KKAAKPKKAASKAPSKKPKATPVKKPAATPKKAKKPKTVKAKPVKASKPKKAKPVK
```

Fig. 6.6 Histone H1 (amino acids 120–180). Amino acid sequence alignment, using the one-letter code, of histone protein H1 for five species. Mismatches (blue) show altered amino acids due to mutations. The high sequence identity among these five proteins shows that they must perform the same function and that all are histones

compare an entire genome sequence between species, this has become a fairly accurate determination of the time elapsed since they diverged, and also of how closely related they are.

Figure 6.7 separates amino acids into three groups, based on how soluble they are in water. Remember that almost all proteins are in an aqueous environment. External amino acids are *hydrophilic*—they love water and therefore, as the protein folds, these amino acids are energetically more stable if they remain on the outside of the folded protein where they can make bonds with water. Internal amino acids are *hydrophobic*—they hate water. As the protein folds, they want to be on the inside of the final structure, as much away from water as possible. The use of the common word *want* is a simple way of explaining that the hydrophobic amino acids are energetically more stable away from water, etc.

Ambivalent amino acids are somewhat flexible so that they can be partially buried inside the protein, while still having part of their structure poking out on the surface.

To summarize: the five hydrophobic amino acids, by their positions along the protein's amino acid sequence, are energetically more stable if buried and so require the protein to fold so as to be in the interior. This determines most of the final structure. The seven external amino acids are also important because they really want to be on the protein's outside, or surface.

The eight ambivalent amino acids then give flexibility in folding. It is the side chains of the external and ambivalent amino acids that participate in chemical reactions. It is important to be aware that these side chains play important roles in:

1) Determining protein structure
2) Catalyzing enzymatic reactions

Without showing actual sequence alignments, Fig. 6.8 displays different organisms and also how closely they are related to the average *Homo sapiens*.

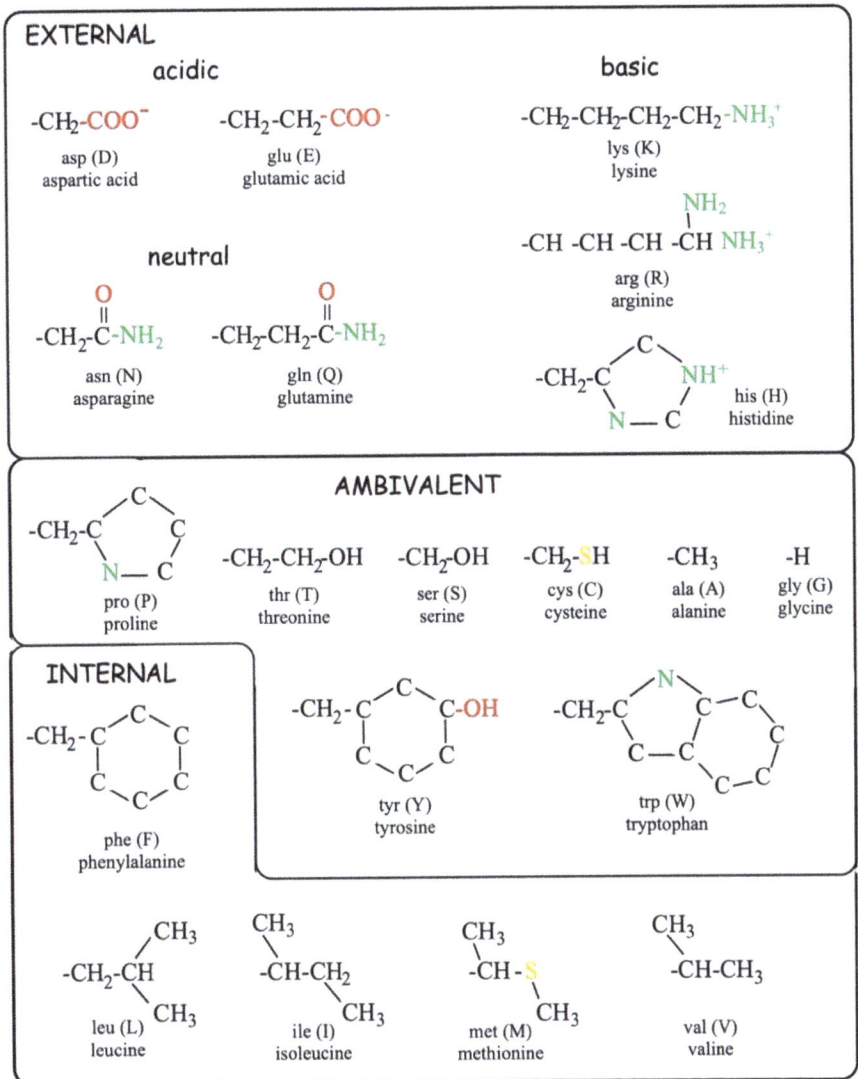

Fig. 6.7 Structures of the 20 amino acid side chains and where their chemical properties position them in the folded protein structure. For each, the horizontal line at the left is the bond to the alpha carbon of the amino acid backbone (Fig. 4.5)

"Average" is a necessary description because, at the level of their DNA, all humans are not absolutely identical. If that were the case, then DNA fingerprinting would have no purpose.

From an evolutionary point, we are a long way from *E. coli* (Fig. 6.8). But yeast cells are eukaryotic and have membrane-enclosed nuclei, mitochondria,

	% common
Human (*Homo sapiens*) 22,268 genes	99.9
Chimpanzee (*Pan troglodytes*) ~22,000 gernes	99
Mouse (*Mus musculus*) ~22,000 genes	90
Mustard (*A. thaliana*) ~25,000 genes	26
Yeast (*S. cerevisiae*) 6,275 genes	23
Bacteria (*E. coli*) ~4800 genes	7

Fig. 6.8 Degree of relatedness of different species

etc., and so we share almost one-fourth of our DNA with yeast. In comparison with any mammal, we share at least 90%; after all, mammals arose fairly recently (Table 4.2). But even with other humans, we are not absolutely identical, as described in the next chapter.

2 Proteins Vary in Size and Complexity

The final type of protein structure is the *quaternary structure*. This refers to those proteins that function only when they are combined with one or more other proteins. Once two or more proteins form a complex, each member protein is called a *subunit*. Most commonly, proteins combine with their own special type. When two such subunits are together, it is called a dimer; three make a trimer, etc.

Table 6.2 lists a few normal proteins to display how proteins vary in size and in their ability to form complexes with other proteins. To appreciate these

Table 6.2 Proteins with subunits

Protein	Subunits Mol. Wt.	No.	Total Mol. Wt.
Insulin	5,733	2	11,000
Hemoglobin	16,500	4	66,000
Lactate dehydrogenase	35,000	4	140,000
Hexokinase	48,000	2	96,000
Phosphofructokinase	83,000	4	336,000
Glycogen phosphorylase	97,400	2	195,000

sizes, which show the protein mass in daltons (which is the weight of a single hydrogen atom), remember that water has a molecular weight of 18 daltons, glucose is 90 daltons, and the average size for a single amino acid is 110 daltons. For simplicity, most of the molecular weights have been rounded to the nearest hundred. Almost 90% of proteins form multimers, and as shown in Table 6.2, dimers and tetramers are the most common.

Most proteins are also large enough that they contain separate folding portions that remain close together after the entire protein has folded, and these separate portions are *domains*. Figure 6.9 depicts two acrobats in an unusual pose or orientation. We easily recognize certain fixed portions of the body: arms, legs, head, etc., even when they are in somewhat unusual positions. If these women were proteins, we would speak of different conformations, resulting from repositioning the different domains.

Almost all proteins larger than 30 kilodaltons have at least two domains. The diversity of protein structures is illustrated in Fig. 6.10. The smallest units of structure that have a folded conformation are modules, which are mostly coded by a single exon in DNA (described in Chap. 9).

Domains normally contain three or more modules. Then a small protein would normally consist of one domain, and larger proteins have two or more domains. This provides larger proteins with more binding sites to be used for a second substrate or for binding inhibitors or activators.

In the next chapter, we will see that domains are normally coded by a fixed portion in some gene, and, by the process of recombination, such a portion of DNA sequence can then be transferred and inserted into some other gene. If, by chance, it is inserted in such a fashion as not to change that gene's reading frame, and also so as to be in the right position and orientation in the folded protein coded for by that gene, then such a domain can find multiple uses in many different proteins.

Fig. 6.9 An example of domains, as in proteins. Shown are acrobats in an unusual position (conformation). Each distinct body portion, such as a head or arm, would be comparable to a domain. (Source: The Austin Chronicle, May 2, 2003)

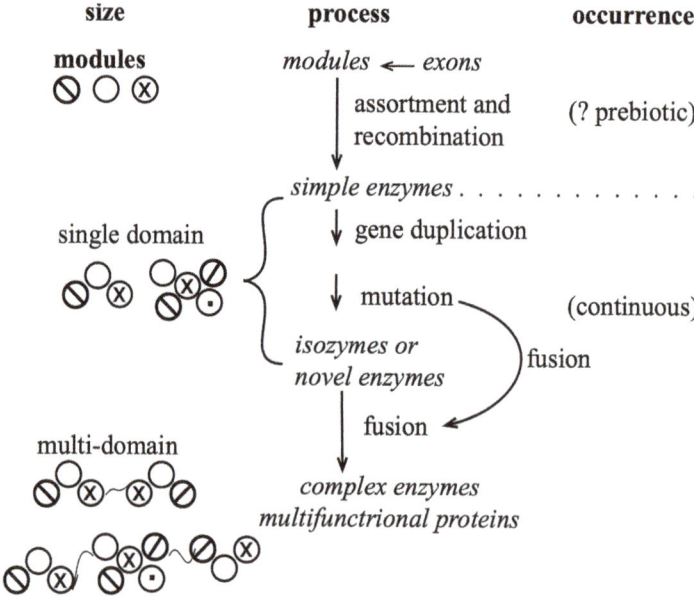

Fig. 6.10 Recombination of exon-coded modules, or larger DNA segments, leads to various larger proteins

3 How Information Is Positioned on a New Protein

The earliest proteins folded mainly as a consequence of having to bury their hydrophobic amino acids. This would result in a stable tertiary structure (Fig. 6.1). Over evolutionary time, additional features could then be added without disturbing the conformation of the original protein, because this conformational shape must be retained to form the catalytic pocket where enzymes bind their substrates and perform some appropriate chemistry. As shown in Fig. 6.11, more than 90% of the sequence is needed simply to guide the folding process into the desired final structure—the major region defining the protein's structure and function (S/F).

Along this portion, after the protein has folded, are positions where the protein can be modified by the addition of chemical adducts, such as phosphate groups, sugar groups, acetyl groups, etc. These modifications enable the protein to have various and appropriate extra features. Scientists universally depict proteins as shown in this figure, with the amino terminus (N) always at the left, and the carboxyl terminus (C) always at the right.

New proteins are always synthesized in the cytoplasm because that is where the ribosomes are (Fig. 2.2). More than half of newly synthesized proteins will remain in the cytoplasm because they are involved with the various branches of metabolism. But for proteins intended to function somewhere else, an initial challenge is to direct them to their final destination, and this always

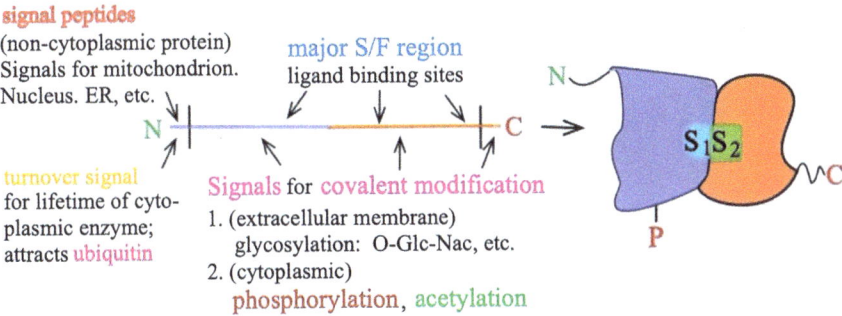

Fig. 6.11 An extended linear protein chain with positions for extra processes. A protein with two domains is depicted, where often each domain separately binds one of the two substrates to form the complete active site. The benefit of such an arrangement may be that if phosphorylation of the blue domain causes it to move slightly relative to the second domain, the active site, formed by S_1 and S_2 - for the substrates, will no longer be complete, and the enzyme becomes inactive. The alternative process is equally possible

involves crossing a membrane into an organelle or out of the cell (Fig. 6.12). This is accomplished with a short amino acid sequence at the N-terminus, often called *signal peptides*, although they are part of the complete protein. These function like a cellular ZIP code to ensure that proteins reach their proper destination.

To appreciate how these become useful, note that the N-terminal and C-terminal parts of a protein do not normally fold into or onto the folded protein. They tend to extend outward, as depicted in Fig. 6.11. Because of this feature, these signal peptide sequences can easily interact with transporters that occur on the different types of organelle membranes, and these transporters only recognize the correct signal sequence for their organelle. This is a remarkably effective sorting system.

The extended linear polypeptide is diagrammed simply to facilitate showing where chemical modifications on the protein may occur. But it should be evident that these changes are only performed on the folded protein and are therefore always on the protein's exterior.

Beginning at the N-terminus:

Noncytoplasmic proteins: Proteins that must cross a membrane, either to exit the cell (extracellular proteins) or to be in an organelle (nucleus,

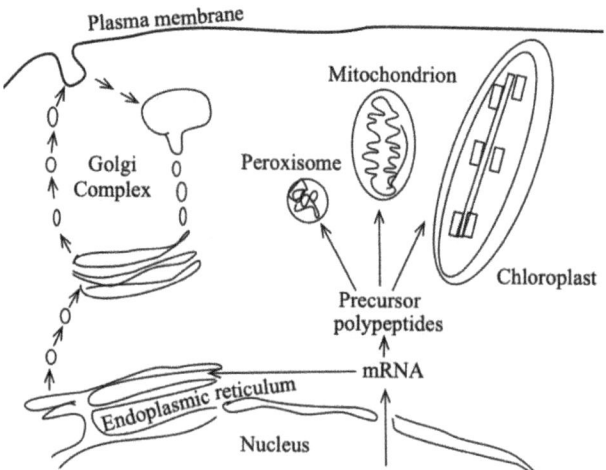

Fig. 6.12 Membrane-bound organelles into which new proteins may be transported. Not all cells have every organelle depicted. "Traffic signals" help proteins reach their destination. Some mRNAs will attach to the endoplasmic reticulum membrane (horizontal arrow) if the protein that they are synthesizing has the appropriate leader sequence for direct extrusion of that protein through the membrane

mitochondrion, etc.), have a *signal sequence* to guide them to the correct membrane for transport through that membrane (Fig. 6.12).

Cytoplasmic proteins: These have a *turnover signal* sequence very near the N-terminus that attracts a proteasome by having ubiquitin attached.

Most proteins are not intended to exist forever. They are synthesized when the cell needs them, and then taken apart by *proteases* via hydrolysis when that use is completed, and the amino acids are recycled. Even if a protein's function is required continuously, some of them are recycled in this fashion in an ongoing process. The proteases are actually part of a protein complex, named the *proteasome*, which is attracted to and binds *ubiquitins*. Ubiquitin is a small protein that serves like a red flag to signal that the protein to which it is attached is damaged or has become superfluous and needs to be destroyed. Along most of the protein sequence are potential sites for the attachment of special compounds:

Proteins that will embed in the extracellular membrane: These frequently have various chemical compounds covalently attached to aid in maintaining their position and in some functions. Such chemical adducts include glycosylation (attachment of glucose), acetylation (attachment of an acetyl group: $CH_3\text{-}COO^-$), or larger fatty acids.

Proteins that will remain in the cytoplasm: The most common attachments are phosphorylation or acetylation. Such chemical adducts mainly function to change the activity of enzymes. Their addition adds negative charge at this position and usually causes the domain or entire protein to change its shape enough to either form or disrupt the active site. This becomes a simple mechanism for regulating enzyme activity, turning the enzyme on or off.

Embedding proteins in a membrane occurs mainly for transporters and receptors. These proteins go from inside the cell, connecting to the cytoplasm, cross through the membrane, and have a portion on the outside of the cell. These consist of two major categories:

Transport proteins: As the name implies, their function is to bind a particular metabolite, glucose as an example, and then momentarily form a large enough pore or opening in the membrane so that the glucose, which is in your blood after the last meal, can pass through the membrane into the cytoplasm, so that cells can use it.

Receptor proteins: These proteins are mostly on the outside of the cell but have a portion that spans the membrane and connects with other proteins in the cytoplasm. An example of this will be shown in Fig. 8.9 with the GPCR (**G-p**rotein-**c**oupled **r**eceptor), which responds to some signal, such as a hormone, and then activates the release, on the internal side, of GEF (guanine nucleotide exchange factor), causing the G-protein to release the GDP, which was keeping it inactive, and then bind GTP to become an active kinase again.

4 Bonds that Hold Proteins in a Stable Conformation

Intracellular proteins are held in a folded conformation that must be the most stable possible. The reason that such structures are so stable is that, if the protein folds appropriately, it will have the maximum number of weak bonds (illustrated in Fig. 6.13), resulting from positioning as many amino acids as possible in an orientation where they can make one or more of these weak bonds.

While these bonds are fairly weak, there are many such bonds, normally at least one for each amino acid (Fig. 6.13). The strength of such bonds, or the bond energy, is shown in Table 6.4. Clearly, covalent bonds are very much stronger. There are four different types of noncovalent bonds, and there are so

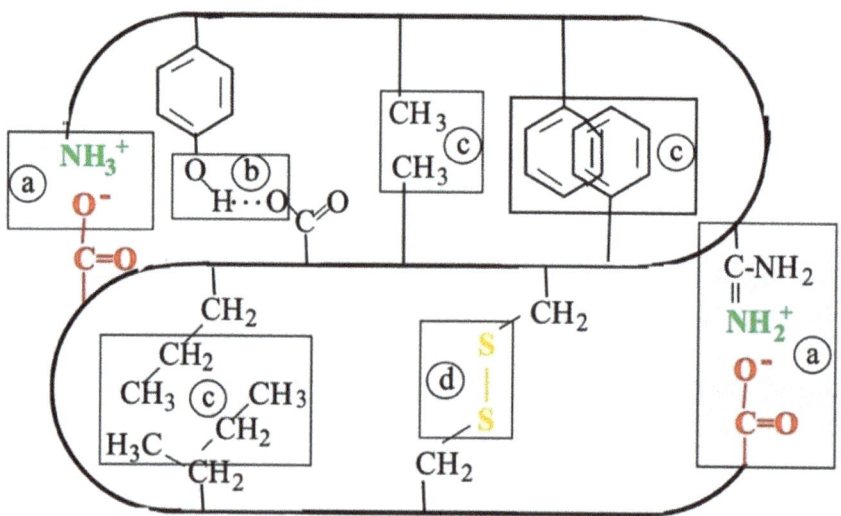

Fig. 6.13 The types of bonds in protein structures. The solid black line represents the backbone of the amino acids, and only the side chains are shown for a few. Disulfide bonds occur only in extracellular proteins

Table 6.4 Bond energies (kcal/mol)

Covalent		Noncovalent	
C-H	98	H-bond	1–5
C-N (peptide)	65	Ionic bond (electrostatic)	1–5
C-C (ethane)	80	Hydrophobic association	3–7
S-S (disulfide; extracellular only)	50	van der Waals	1–2

many more of them that they determine how a chain of amino acids will fold into the protein structure that it will form.

The important bonds that determine structure are:

(a) *Ionic bonds*: At a physiological pH (about 7.4 for mammals), amino groups are positively charged, and carboxyl groups are negative. The attraction of these two groups with opposite charges forms an ionic bond (Fig. 6.13a).
(b) *Hydrogen bonds*: The hydrogen atom on hydroxyl groups (–OH) is normally shared with an adjacent oxygen (–OH···O) if the second oxygen is close enough (Fig. 6.13b).
(c) *Hydrophobic interactions*: These are not truly bonds, but when two hydrophobic groups are next to each other, they both avoid water and are therefore more stable. Note in Fig. 6.13 that there are many possible types of hydrophobic groups.
(d) *Disulfide bonds*: These bonds are covalent (strong) (Fig. 6.13d) and are found only in extracellular proteins. Antibodies are examples.

5 Synthesis of Proteins

As will be shown in Fig. 9.11, the mRNAs, after being synthesized in the nucleus, where the DNA is, move out into the cytoplasm, where the ribosomes are, and are then translated by the ribosomes into new proteins.

Proteins are linear polymers, as diagrammed (Fig. 6.14). This diagram starts with the first nucleotide at the 5′-end, but this is not where translation starts. To start this process, initiation factors, which are special proteins that recognize a sequence on the mRNA that is far to the left of the 5′-end shown, bind there and then attract ribosomes to bind there. The ribosome is designed to slide along the mRNA toward the 3′-end until it encounters the first start codon, AUG. It pauses here long enough to permit a tRNA (transfer RNA,

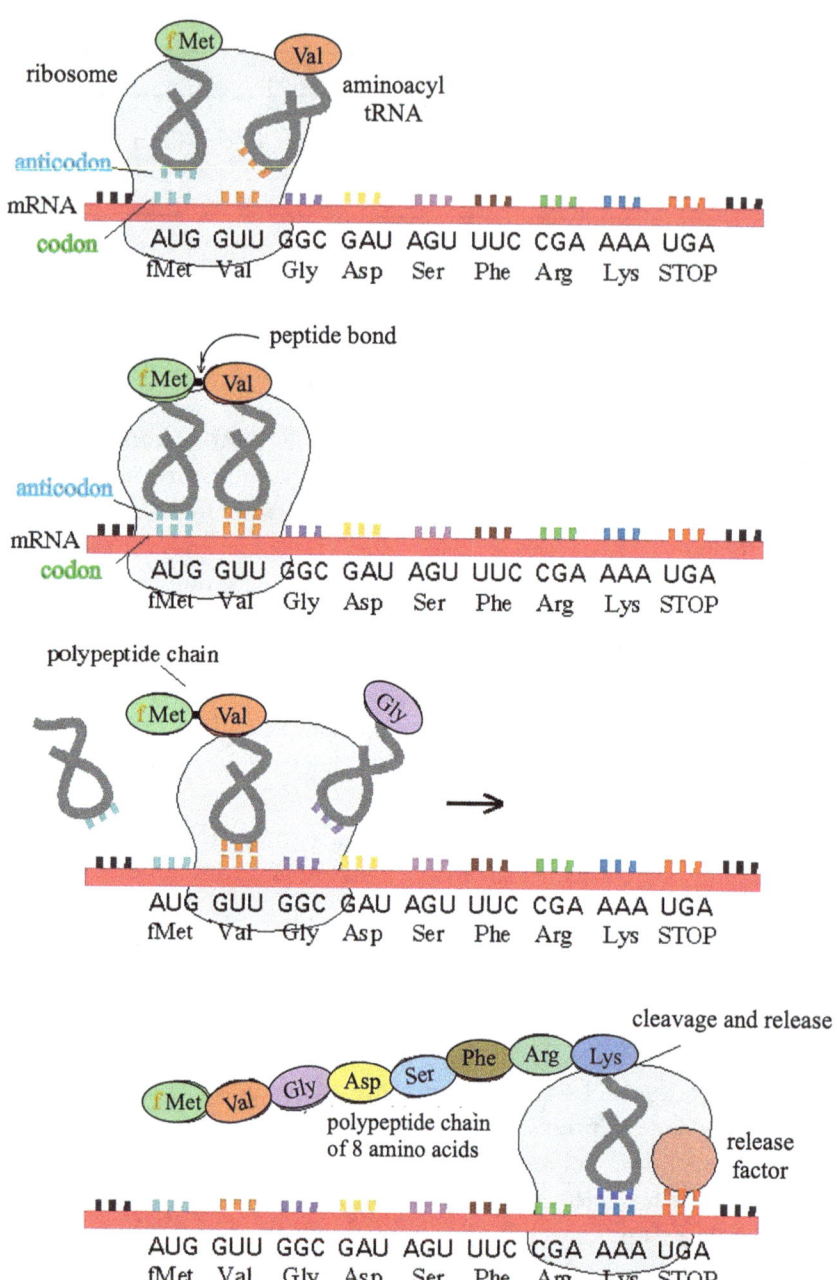

Fig. 6.14 Initial stages in protein synthesis in bacteria. After being attracted by initiation factors, the ribosome binds to the mRNA and begins translating the message, starting at the first AUG codon. Codons on the mRNA (in capital letters) are also shown by three short vertical bars, and the same for anticodons on the tRNA. Below the codons are the amino acids that they code for

because it transfers amino acids to the end of the growing protein chain), with the correct anticodon to line up and base-pair with the codon on the messenger RNA. The ribosome is large enough to hold an incoming second tRNA if it can base-pair (Fig. 6.14) at the second codon. At this point, the ribosome catalyzes the bond formation between the oxygen atom at the end of the first amino acid and the nitrogen atom at the start of the second amino acid (shown by a black dash between the amino acids). This is a *peptide bond*. And because proteins have many amino acids with such bonds, they may be called *polypeptides,* although this term is normally restricted to fairly small proteins containing less than 60 or 70 amino acids.

Because the ribosome starts translation at the AUG codon, which codes for the amino acid methionine, all proteins have this amino acid at their beginning. Once the peptide bond is formed between the first two amino acids, the ribosome slides one codon over along the mRNA. The first tRNA is no longer stabilized by the ribosome and dissociates. When the ribosome is at the third codon, a new tRNA can bind, and its attached amino acid is then joined to the initial dipeptide, making a tripeptide. This continues until the ribosome reaches the STOP codon. There is no tRNA that has an anticodon for binding here, but a special *release factor* can bind here and bring translation to a stop because the release factor has no attached amino acid. For simplicity, this protein stops at 8 amino acids (Fig. 6.14), but readers already know that most proteins contain more than 200 amino acids.

Resources

Review of protein structure by Jane Richardson: https://pubmed.ncbi.nlm.nih.gov/7020376/

Review of protein biosynthesis: https://en.wikipedia.org/wiki/Protein_biosynthesis

Noncovalent interactions in proteins: https://pubs.acs.org/doi/10.1021/acsomega.3c00205#

7

Our Amazing Hemoglobin: Blood and the Circulatory System

Abstract Brief description of the human circulatory system. Functions of hemoglobin in delivering oxygen and removing waste products such as acid protons and carbon dioxide. The brain's unusual requirement for 20% of our blood flow to get enough oxygen. Explanation of hematocrit. Description of the many types of globin proteins. Explanation of using units of torr for the gas pressure, for concentrations of oxygen, and pO_2 as the binding constant of hemes for oxygen. Explanation of how binding a compound by any protein, such as an enzyme or hemoglobin, is an on–off process. Normally, hemoglobins release one half of their bound oxygen molecules in the capillaries to be used by adjacent tissue cells. Explanation of why the different globin proteins must have different affinities for oxygen to achieve the net movement of oxygen from the lungs to the body's tissues. Oxygen-binding curves by hemoglobin are sigmoid, indicating that the hemoglobin is changing its affinity for oxygen in response to the surrounding concentration of oxygen, acidity, or temperature. These allosteric effectors signal a greater need for oxygen. Fetal hemoglobin must bind oxygen more tightly than adult hemoglobin.

Keywords Hemoglobin • Circulatory system • Blood vessels • Lungs • Capillary • Heart • Aorta • Arteries • Veins • Oxygen • Carbon dioxide • Acid protons • Erythrocytes • Lymphocytes • Platelets • Macrophages • Hematocrit • Myoglobin • Cytoglobin • Neuroglobin • p50 and pO_2 • Allosteric • Sigmoid • Placenta

Our circulatory system is an amazing organ, although almost no anatomy or biology book would refer to it as an organ. Like more familiar organs, our circulatory system has a highly defined structure, with all the blood vessels always connecting the heart and lungs to the rest of the body.

In this simplified scheme (Fig. 7.1), air containing 21% oxygen is inhaled into the lungs and diffuses through the membranes of capillary vessels. These vessels merge into the vena cava and enter the heart, and then the blood is pumped via the aorta to the body. In Fig. 7.1, the subscript numbers on oxygen molecules denote how many oxygen atoms are binding together to form a molecule. When the larger arteries reach the tissues, they branch into smaller vessels and then enter very small capillaries where oxygen is released to the adjacent cells (see Fig. 7.2).

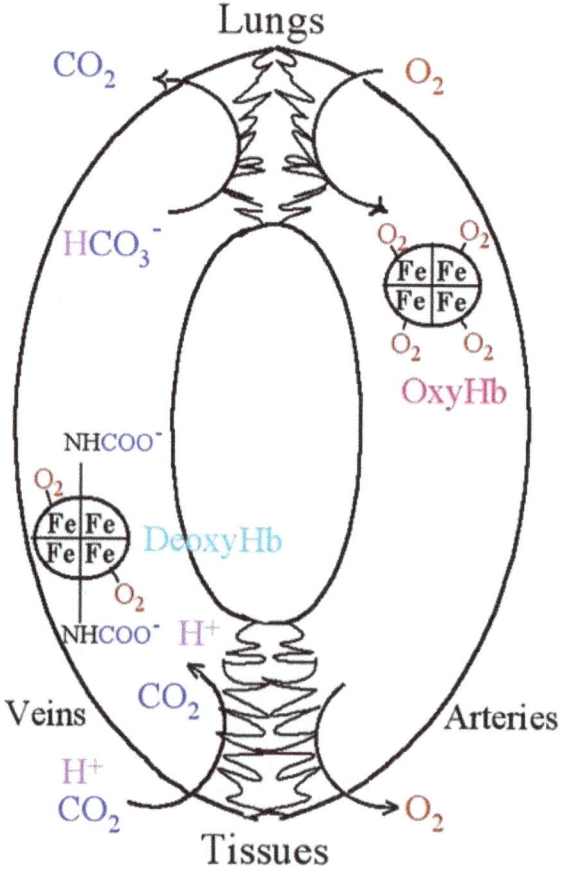

Fig. 7.1 Functions of hemoglobin A as it circulates through the body. O_2 is carried from the lungs to the tissues; CO_2 is carried back to the lungs. HbA in veins is still half full with oxygen

7 Our Amazing Hemoglobin: Blood and the Circulatory System

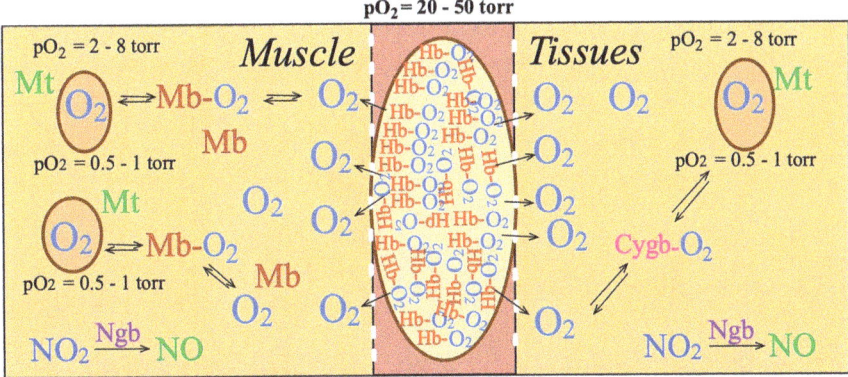

Fig. 7.2 Passage of a red blood cell through a capillary. The red blood cell is completely full with HbA molecules binding O_2. Adjacent cells contain Mb in muscle, or Cygb and Ngb in other tissues. Note the changing concentrations of oxygen (pO_2) in blood, above the capillary, in the cells, and at the mitochondria, where oxygen is constantly being used. *Hb* hemoglobin A, *Mb* myoglobin, *Cygb* cytoglobin, *Ngb* neuroglobin, *Mt* mitochondrion

Blood vessels are the major transport system for the body. Not only do they supply the constantly needed oxygen, but all nutrients, after absorption in our intestine, are carried via the blood to the cells. Blood also helps in the disposal of waste products. Carbon dioxide (CO_2) is produced by oxidizing fuel molecules for energy (see Chap. 11), and this waste product is carried in the veins to the lungs and exhaled.

Here is an interesting fact to explain how important blood is to the brain. In a normal adult, the brain weighs about 1.4 kg (~3.1 lbs.). For a normal adult with a body weight of 75 kg (165 lbs.), this would be about 2% of the total body weight. However, the brain normally gets about 20% of the blood flow. The brain is a high-maintenance organ.

In rapid muscle action of any kind, such as exercising, the oxidation of glucose may lead to the formation of lactic acid (discussed in Chap. 11), and the acidic protons (H^+) are then also removed. One molecule, hemoglobin, has evolved to have so many functions. To appreciate this, imagine that you have ordered some pizza. After the pizza has been brought to your house, the delivery person will also take away your empty pizza cartons and dirty napkins from the last order.

Blood carries these types of important cells:

Red blood cells (erythrocytes).
White blood cells (lymphocytes).
Platelets.

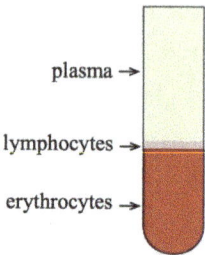

Fig. 7.3 Components of a blood sample

White blood cells are important components of our immune system and include several types of different cells. Among these are the *macrophages* (Greek: big eaters) that are able to engulf and devour smaller invading agents such as bacteria and viruses. Platelets are the smallest cells in a human (see Table 2.2). They have become so small by not retaining their nuclei or mitochondria, which normally take up more than one-half of a cell's volume. Being small, they can aggregate with each other to make a plug of any shape to fill an injury site where blood is leaking out.

However, because red blood cells are so important, clinical laboratories normally measure a person's *hematocrit*, the volume of our blood that is formed by the red cells. As shown (Fig. 7.3) in a test tube, after centrifugation, erythrocytes constitute almost 50% of our blood. For normal, healthy adults, the hematocrit for females is 36–44%, and for males it is 41–51%. That is why that old saying is correct: blood is thicker than water.

1 We Have Many Globin Proteins

Globins are the proteins that carry oxygen through the blood and also within cells. Although they are not actually enzymes, they have so many features in common with enzymes and are among the best-studied proteins because it is so easy to get samples of blood in which the erythrocytes are almost half of the cells (Fig. 7.3) and are almost completely filled with hemoglobin. Therefore, hemoglobin was the first protein to be purified and studied so extensively. Like enzymes, globins have a binding site where a special ligand, oxygen, binds. Unlike enzymes, most globins perform no chemistry on this ligand.

Because globins are so vital in providing oxygen constantly to all cells, where oxygen is mostly used by the mitochondria in the formation of ATP (described in Chap. 11), many different types of globins have evolved (Table 7.1). Note that here "globin" is the generic name for these various

7 Our Amazing Hemoglobin: Blood and the Circulatory System

Table 7.1 The different oxygen-carrying globins

Globin type	Oligomer	Location	Carries O_2 between	p50 (torr)
In blood (erythrocytes):				
Hemoglobins:				
HbA	$\alpha_2\beta_2$	Adult erythrocytes	Lungs → tissues	27
HbF	$\alpha_2\gamma_2$	Fetal erythrocytes	Lungs → tissues	7
In tissue cells:				
Myoglobin, Mb	α	Muscle	Cell membrane → mitochondria	0.9
Cytoglobin, Cygb	α_2	Brain, peripheral tissues	Cell membrane → mitochondria	5.4
Neuroglobin, Ngb	α	Brain, peripheral tissues	Acts as nitric oxide (NO) synthase	7.5

proteins. Two distinct features make each globin unique. The first is their location. The second is how tightly they bind oxygen.

In Table 7.1, the subscript number in the oligomer column refers to the normal structure of that protein. HbA (A for adult) contains four hemoglobin subunits (Fig. 7.4), where α and β denote that there are two types of hemoglobin subunits. As shown by its name, Hb F (F for fetal) occurs mainly in the developing fetus, and within the first year after birth, it is steadily replaced by HbA.

The final column in this table shows the affinity (p50) for the oxygen that it will carry. In Chap. 1, we saw the affinity of an enzyme for its substrate defined as K_M, with concentrations in micromolar or millimolar, because the substrate molecules are in a liquid solution, such as a test tube or a cell. Oxygen is a gas, and the amount of gas molecules in a solution varies with the pressure of that gas above the solution.

It has already been noted that oxygen comprises 21% of normal air at sea level, because oxygen concentrations in air decline at higher altitudes. People can become anoxic at about 15,000 to 16,000 feet above sea level (above 5,000 meters).

Definitions

pO_2 atmospheric pressure of oxygen (= concentration); units of torr (= mm Hg).

p50 concentration of oxygen at which 50% of a globin solution is bound with O_2

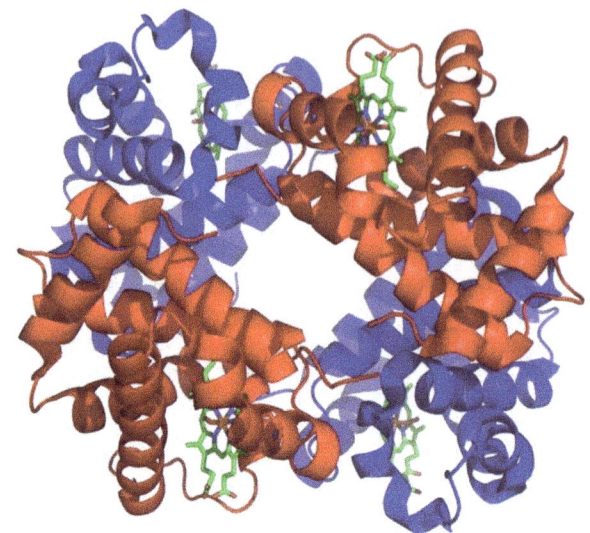

Fig. 7.4 Three-dimensional structure of hemoglobin. α/β dimers are blue and red. The light green molecules represent heme rings. (Source: Richard Wheeler, Wikipedia)

The units for pressure of gases are then measured in torrs, named for the Italian scientist who first described these and also invented the barometer. The amount of air, or atmospheric pressure, at sea level is referred to as one atmosphere. But with his new instrument, Torricelli measured this as being 760 torr. The concentration of oxygen is defined as the *partial pressure*, written as pO_2, and is then $0.21 \times 760 = 160$ torr.

We need to consider one more aspect of how proteins, whether enzymes or globins, bind some molecule. While for enzymes these are normally defined by S, for substrate, for nonenzyme proteins, such as the globins, we will use the more general symbol L, standing for *ligand* (from *ligare*, the Latin for binding).

As explained in Chap. 1 (and see Fig. 1.1), for a protein to bind a ligand, the ligand must bump into the protein, and more specifically, it must connect to the protein's exact binding site. The affinity constant, whether K_M or p50, always defines the amount of that ligand needed for 50% of the proteins to bind it. Therefore, if the K_M or p50 has a low value, it means that only a low amount of these ligands is needed because they bind so well. As the p50 value becomes larger, globins are binding oxygen more weakly.

It must also be remembered that binding of such ligands by an enzyme or hemoglobin is an on-off process. When humans pick up an apple or a ball, they do not normally drop it. They can hold any object for quite a long time, if it is not too heavy or too big. Enzymes or globins are not rigid; they are slightly flexible, and the pocket where a ligand binds can flex and open a little, so that the ligand can diffuse away. Now, if two similar but different types of proteins are in the same solution, and there is a limiting concentration of ligands to bind to them, the protein type that binds the ligand more tightly (lower p50 in Table 7.1) will bind more of them.

This does not mean that a single protein of that type will bind more ligands; they usually only bind one. It means that a population of this protein type will have more of its members binding the ligand. This becomes an important feature when the carrier proteins work between two regions in the body with distinctly different concentrations of a ligand, oxygen for example, in our bodies (Fig. 7.5). The oxygen pressure in the lungs (pO_2) is normally above 130 torr. While in the outside air the pO_2 is 160 torr, the lungs never completely empty the remaining air after the last breath, still containing some CO_2, so the lung pO_2 is this much lower than the external air. HbA has two different p50 values (explained in more detail in the next chapter), as shown in Fig. 8.2, that have evolved to be ideally suited for binding oxygen in the lungs so that it becomes almost completely filled, and then releases this oxygen in the capillaries where it is picked up by myoglobin or cytoglobin with p50s that bind much more tightly than HbA (Fig. 8.2).

This exchange of oxygen across the capillary membranes is diagrammed in Fig. 7.2. Because the HbA molecules have a p50 in the lungs of about 10 torr, they will normally fill up completely. When HbA reaches the capillaries, where the oxygen concentration is much lower, it will change to a much weaker p50 of about 300 torr and therefore release at least half of the bound oxygens that they are carrying (see Fig. 7.6), which can then diffuse into the adjacent cells, where they are immediately bound by Mb or Cygb, which have lower p50s, meaning tighter binding of oxygen. Inside the cells, these globins will then diffuse toward the mitochondria, where the local concentration of O_2 is so low, due to their constant use in making ATP, that the cellular globins will also unload most of their oxygen (see Fig. 7.2).

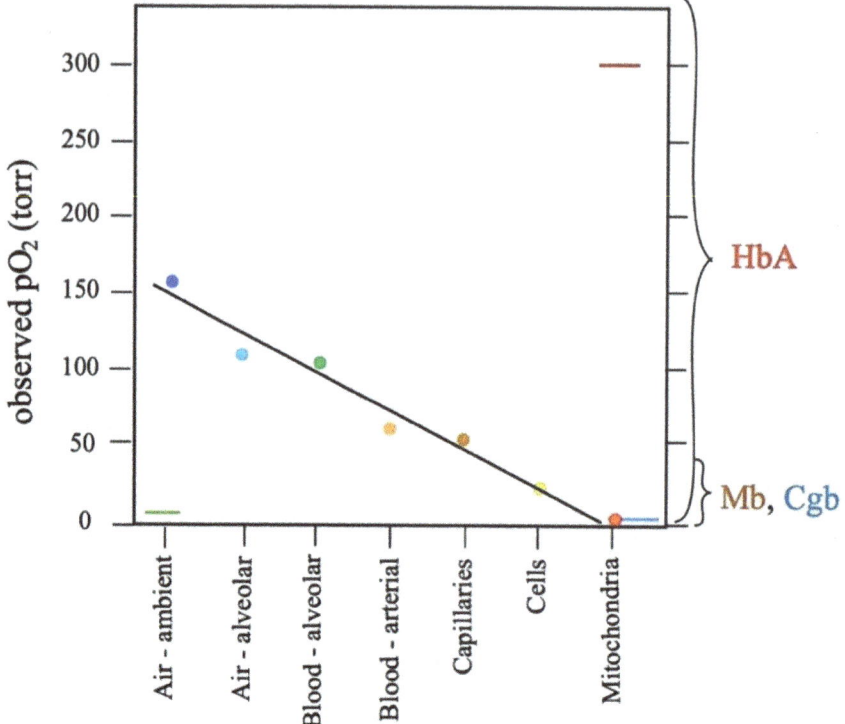

Fig. 7.5 Change in oxygen concentration (pO$_2$) along the path that oxygen is transported from the lungs to the cells' mitochondria. The red and green horizontal lines (top right and bottom left) represent the p50s for HbA at these different conditions of pO$_2$; the blue horizontal line represents the p50 for Mb or Cgb. The brackets include the oxygen concentrations at which each globin can bind or release oxygen

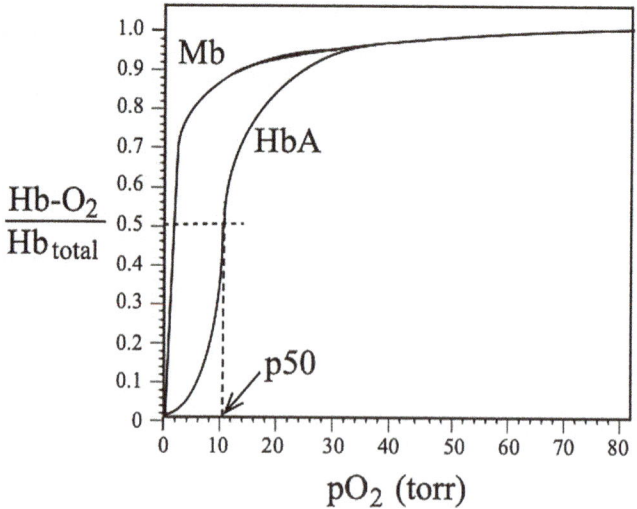

Fig. 7.6 Binding curves for oxygen by adult hemoglobin and myoglobin. The vertical lines for each binding curve show the amount of oxygen necessary for 50% loading. *HbA* hemoglobin without effectors, *Mb* myoglobin

2 Having Different p50 Values Is Important for Globin Functions

If all globins had the same affinity for oxygen (p50), there would be no possible exchange of oxygen between the different globins. HbA has the weakest affinity overall. When it goes through the capillaries of the lungs, the available oxygen concentration is quite high, so that even with a modest affinity of 10 torr, it loads up almost completely. When it returns to the capillaries in any tissue, where the local oxygen concentration is between 20 and 50 torr, HbA will unload almost 50%. This may seem difficult, but HbA is a tetramer (Fig. 7.4), meaning that it is composed of four subunits ($\alpha_2\beta_2$), and so only two of the four subunits will release the oxygen that they are holding.

3 Oxygen Binding Curves

A normal binding, or loading, curve for oxygen by two different globins is shown in Fig. 7.6. When we look at graphs such as this one, it is fairly standard for the viewer to begin at the lower left corner of the graph and then visually follow the curve upward to the top right. But one can equally well begin at the top right and follow the curve toward the bottom left corner. Then one would see an unloading curve. And this is a better perspective for understanding how both HbA and Mb molecules bind oxygen at higher pO_2, always more to the right on a graph, and then release oxygen at lower pO_2, always toward the left. The amount of binding or loading is shown as $\frac{Hb-O2}{Hb\ total}$. The scale on the left axis then goes from 0 (empty) to 1.0 (totally full). The horizontal dotted line shows the position for 50% loading, which is also the surrounding oxygen concentration at which each globin unloads enough to be half empty.

Readers who have ever had an oxygen sensor placed on their fingertip at a medical clinic already know that, for a healthy adult, the sensor should measure oxygen concentration at 98%, or perhaps 97%. This clearly shows that blood coming to the capillaries, before unloading oxygen, is still almost completely full. This is what we see for HbA at the top right in Figs. 7.6, 7.7, 7.8, 7.9, and 7.10.

In Fig. 7.6, HbA binds oxygen somewhat moderately, and it will be almost completely empty at a pO_2 where Mb will be half full. Readers may see this as inefficient, but if there are enough HbA molecules in the red blood cells and

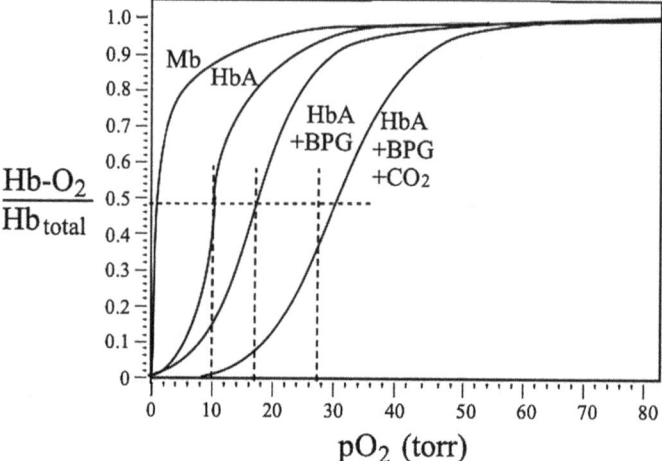

Fig. 7.7 Effectors change the affinity of HbA for oxygen. The vertical lines for each binding curve show the amount of oxygen necessary for 50% loading. *HbA* hemoglobin without effectors, *Mb* myoglobin. *BPG* bisphosphoglycerate

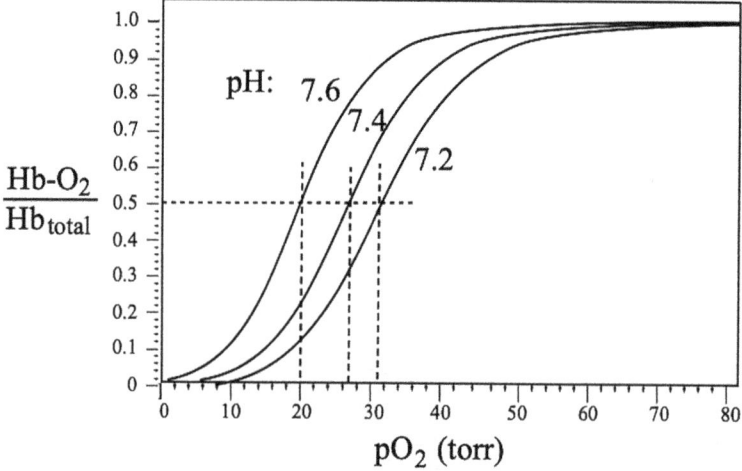

Fig. 7.8 Increased acidity (lower pH) increases the release of oxygen

enough Mb molecules in every muscle cell, then even with only half of the oxygen being unloaded by each HbA, the overall transport of oxygen to the cell's mitochondria will be enough for healthy mammals to be very active.

We must understand that Mb has also evolved to load almost fully at the pO_2 of the capillaries, and then it must unload at the pO_2 of the

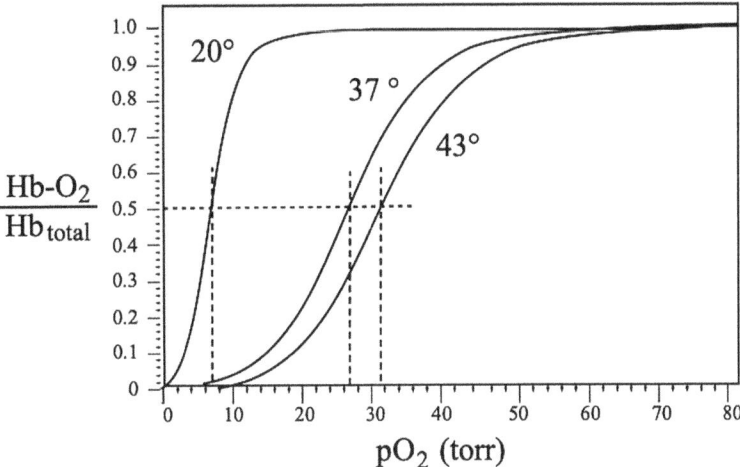

Fig. 7.9 Increased body temperature increases the release of oxygen by HbA. Temperatures shown are in centigrade: 20° = 68 °F; 37° = 98.6 °F; 43° = 107.6 °F

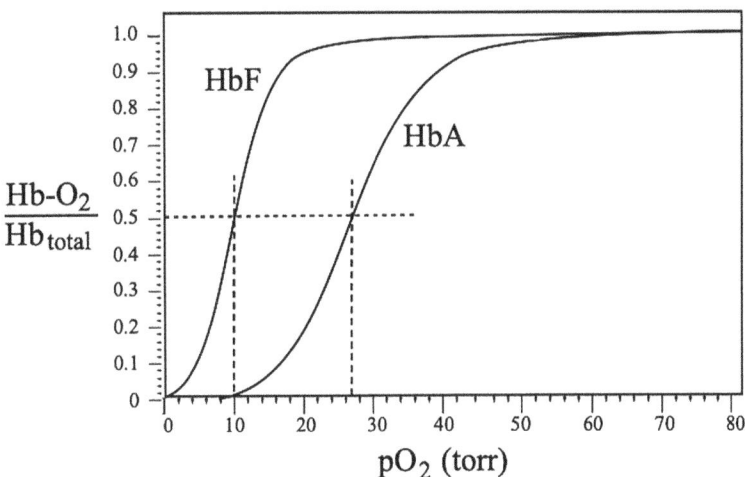

Fig. 7.10 Fetal hemoglobin, HbF, must bind oxygen more tightly than HbA to assure significant transfer of oxygen in the placenta

mitochondria. Fine-tuning the affinity of globins for oxygen occurred over millions of years by having the hemoglobin protein become able to respond to metabolites in our cells that signaled a need for more oxygen delivery. This is demonstrated in Fig. 7.7, where the adult hemoglobin, HbA, has a much stronger binding of oxygen (lower p50) when no effectors are present, just as in Fig. 7.6. When we are consuming sugar molecules to produce energy, bisphosphoglycerate is a normal byproduct that is released into the blood and therefore can act as an indicator of a greater need for oxygen.

By binding to HbA, it causes this hemoglobin to bind oxygen more weakly, and thus release the bound oxygen. In a similar fashion, when oxidation of sugar molecules is rapid, the carbons in these food molecules are oxidized and become carbon dioxide (CO_2), which also binds to HbA and causes it to bind oxygen more weakly, making it easier for oxygen to be released and to enter cells. One can then see how our adult hemoglobin (HbA), with a p50 at about 10 torr, has been adapted to bind these two effector molecules, which cause it to have a weaker affinity with an average p50 for oxygen at the standard 27 torr.

It has already been mentioned that HbA also binds acidic protons (H^+), and that the body, especially muscle tissues, makes more acid as a by-product when oxidizing sugars rapidly during vigorous exercise. These acidic protons make blood more acidic, as normally indicated by the pH of the blood. Figure 7.8 shows oxygen binding/unloading curves in response to the acidity of the blood. A pH of 7.4 is normal for a healthy adult, and pH values below 7.1 lead to serious clinical problems and can become lethal. Even a moderate or tolerable change in acidity (pH 7.2) shifts the binding curve to the right, indicating weaker binding of oxygen, and therefore better release of oxygen to the adjacent tissues. A comparable and opposite change occurs if the blood becomes more alkaline (pH 7.6), but this is metabolically most unlikely.

HbA also responds to fevers. A fever is a very primitive response by the body to fight an infection, and it may sometimes be helpful because some bacteria begin to die if the body temperature goes above about 104° (40 °C). To maintain a fever, the entire body must be at this higher temperature, and this requires the "burning" of body fuels, more accurately, the oxidation of fats and sugars. To enable the body to oxidize more fuels requires more oxygen.

4 Fetal Hemoglobin Binds Oxygen More Tightly

By this stage of our discussion of different globins and binding constants, readers should be able to appreciate that the remarkable process whereby the fetus gains oxygen requires another special globin. Maternal and fetal blood never mix directly. They are in separate circulatory systems that come very close to each other in the placenta, where they are separated by a capillary membrane, similar to Fig. 7.2.

As mentioned, with HbA, binding of oxygen is sufficiently moderate that the oxygen dissociates constantly. If the fetal hemoglobin, HbF, on the other side of the capillary membranes binds oxygen more tightly, with a lower p50, then there will be a net transport of oxygen to the fetal circulation.

As illustrated in Fig. 7.10, the affinity of HbF is very similar to that of HbA when it is not binding effectors (see Fig. 7.7). It is very likely that HbF occurred as a gene duplication of HbA, but this gene did not undergo the changes that would make the resulting globin bind effectors that would weaken its binding to oxygen. Therefore, the binding curves for HbF and HbA without effectors look the same. At the normal pO_2 of tissues, where HbA becomes about 50% unloaded, HbF will fill up almost completely. Although HbF binds more tightly, it is still able to release enough oxygen to the growing fetus. After all, the fetus is doing very little muscular work. It requires oxygen mainly to oxidize fuels (sugars and fats) that it also obtains from the maternal circulation. These fuels are then used to make ATP, necessary for the constant biosynthesis of the various molecules needed by all cells as the fetus steadily grows larger.

Without having any nucleus, each erythrocyte has a shortened lifespan of about 100 days. Therefore, every day about 1% of our erythrocytes die and are degraded. How can we afford to constantly replace them? Let's do the math: normal adults have about 4 quarts of blood (~3.8 l) with an average hematocrit (males + females) of 43%. That would be 8 pints, or ~ 4 L (each = 16 ounces), or 128 oz. × 0.43 = 55 oz. (~140 g) of RBC cells. At 1% each day, that is about ½ oz., or 1.4 g, of RBCs each day. We can easily do that.

There has been an ongoing effort to make an artificial blood substitute. This would enable a constant availability of this amazing fluid, which, after being donated, now has a shelf life of about 42 days when stored under ideal conditions. This is even shorter than the lifetime of blood cells within our body. Despite the efforts of many scientists in several different countries, no successful product has emerged to be readily usable.

To help clarify this never-ending need: erythrocytes have evolved to optimize oxygen-carrying capacity by expelling their nuclei, thereby more than doubling their internal cellular volume, and so contain much more HbA. But therefore, they also become short-lived.

Resources

How does the blood circulatory system work? https://www.ncbi.nlm.nih.gov/books/NBK279250/#:~:text=The%20blood%20circulatory%20system%20(cardiovascular,it%20back%20to%20the%20heart

Description of hemoglobin types and allostery: https://en.wikipedia.org/wiki/Hemoglobin

Discussion of red blood cells when they are healthy, and when they are diseased: https://my.clevelandclinic.org/health/body/21691-function-of-red-blood-cells

8

Allosteric Enzymes: Shape Shifters

Abstract Description of how adult hemoglobin (HbA) is allosteric, having a T conformation with poor affinity for oxygen and an R conformation with much higher affinity for oxygen. With a stronger affinity in the lungs, HbA loads up fully with oxygen; with a weaker affinity in the capillaries, HbA will unload at least half of the four bound oxygens. A theoretical globin with the same p50, but no cooperativity, would deliver at best one oxygen. Therefore, cooperativity doubles the amount of oxygen delivered. Hemoglobin by itself is most stable in the T conformation and can change toward the R conformation as oxygen concentrations rise, as in the lungs. Several activators, denoting a greater need for oxygen, also change HbA's affinity for oxygen. HbA exemplifies K-type allosterism, meaning that conformational changes alter the K_M or the p50. The blood clotting enzymes are examples of V-type allosterism, meaning that effectors alter their maximum velocity. The blood clotting cascade is described in detail. V-type clotting enzymes are synthesized as inactive precursors, which can exist in the blood without having any effect. Activation by other enzymes in this cascade results in cleavage and removal of a small part of the original protein, leading to a conformational change to make the remaining protein active. Activation of the G-protein Ras is described as an example of how a protein can be induced to have an active conformation, for a limited time of some minutes, because it slowly hydrolyzes GTP, the activating factor.

Keywords Adult hemoglobin • HbA • Allosteric • Oxygen • R conformation • T conformation • Zymogen • Protease • Blood clotting cascade • G-proteins • G-Protein-Coupled Receptor • Biological time clock

1 Hemoglobin Was the First Allosteric Protein Observed

An important distinction for HbA is its ability to change its affinity for oxygen. Most enzymes, and most globins, have a constant, invariant affinity for the ligand that they normally bind. Such binding or loading curves are always hyperbolic. This means that at the lowest concentration of ligand, the binding increases directly with the increase in ligand concentration. Then, at higher ligand concentrations, the binding curve begins to plateau, as the proteins' binding sites all become occupied. This was exemplified by the curves for myoglobin in the preceding chapter.

But being able to change its affinity is important for HbA because this allows it to have a somewhat stronger affinity (lower p50) when it enters the lungs, and then change to a weaker affinity when it arrives in the tissue capillaries to increase unloading of oxygen. To help in understanding this, we will now examine Fig. 8.1, with two separate oxygen binding curves: one for the normal HbA, and one for a theoretical globin, Gb, that was mathematically produced with an equation for binding.

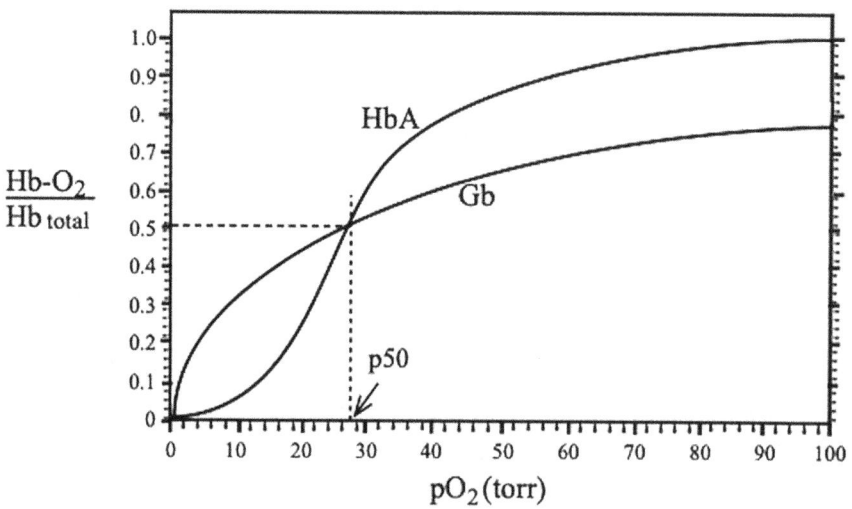

Fig. 8.1 Curve for normal HbA in the presence of effectors, compared to a calculated curve for a simulated globin transporter for oxygen, Gb, with a p50 identical to HbA. The constant-affinity transporter, with a p50 of 27 torr, would not be able to load completely in the lungs and would not unload extensively in tissues. It would deliver about 1 O_2 per tetramer per round trip. By comparison, HbA—with its change in conformation and affinity—can deliver at least 2 O_2 per round trip

The binding, or unloading, curve for HbA is the same as in the graphs in Chap. 7. It has an inflection near the bottom left of the curve. When oxygen concentrations are very low, HbA assumes the T conformation (T for "tense"), the conformation of HbA that binds oxygen poorly and is therefore good for unloading. As oxygen concentrations increase, HbA changes to the R conformation (R for "relaxed") and now binds oxygen more tightly, which is better for loading in the lungs. Because the shape of such a binding curve resembles the letter S, it is usually designated as a *sigmoid* binding curve and is routinely seen as evidence for *allosteric* (from the Greek: *allos* = other; *stereos* = solid) changes, meaning that the protein can have two or more different shapes that produce changes in the enzyme's affinity for a ligand.

The simple hyperbolic shape for Gb (Fig. 8.1) reflects a simple ligand-binding function. The sigmoid (S-shaped) curve for HbA is indicative of interaction, or cooperativity, between the four subunits in the hemoglobin tetramer (Fig. 7.4). Note that neither HbA nor Gb readily gives up (dissociates) all its O_2 until the surrounding O_2 concentration becomes very low. Since Hb is a transporter, additional mechanisms have evolved to facilitate unloading (dissociation) of oxygen.

Because of this change in affinity by HbA for oxygen, the p50 value, normally stated as 27 torr, is actually an average value for the two distinctly different affinities shown in Fig. 8.2. The allosteric transition in HbA actually results in a 30-fold change in affinity for oxygen.

This is a remarkable accomplishment. In terms of the number of ATP required, it costs the cell the same amount to make HbA, Mb, or the potential Gb, as the first two proteins are almost the same size, as measured by the number of amino acids needed to make each protein. But, by being allosteric

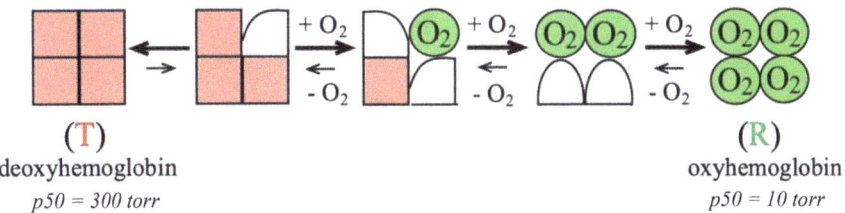

(T)
deoxyhemoglobin
p50 = 300 torr

(R)
oxyhemoglobin
p50 = 10 torr

tissue capillaries: 20 - 50 torr --- actual oxygen concentration -- 110 -120 torr: lung capillaries

Fig. 8.2 A mechanism for allosteric effects. The T conformation occurs when all subunits are empty. This empty conformation may briefly have one subunit approach the R conformation. Binding of one molecule of oxygen to this subunit stabilizes this mixed assembly, which in turn initiates a conformational change in the two neighboring subunits. Binding of oxygen shifts the equilibrium toward the R state. Note that the T and R conformations have different affinities for oxygen

and having the ability to change its affinity for oxygen, HbA can be at least twice as effective in delivering oxygen. All companies would love employees like that.

It is also worth noting that the heavy bold arrows in Fig. 8.2 indicate a favorable change in terms of the free energy. That means that in the absence of oxygen, the HbA tetramer (4 enzyme subunits binding together to make a complex) is most stable in the T state, at far left. But as oxygen becomes more abundant and binds to the hemoglobin subunits, the R conformation becomes energetically more favorable, at far right.

Naturally, in the circulation of a person, hemoglobin is never completely empty; this happens only in laboratory experiments or after death. Within a red blood cell, HbA will always exist as either fully bound, coming out of the lungs, or still having at least one oxygen in the tissue capillaries.

A note on the terminology for HbA conformations: The designations T and R were first proposed by François Jacob and Jaques Monod in their original paper describing allosteric mechanisms. In 1965, the energetics of proteins were not yet that well established, and the definitions that they chose for the inactive state (T for *taut*) and active state (R for *rest*) are opposite to the actual energy levels of these enzymes or proteins with the definitions that we now understand.

As shown in Figure 8.3A, the T state is at a lower energy level than R because T is the normal conformation. R generally occurs only when the protein is binding an activator, A. If T binds an inhibitor, it will be even more stable and less likely to be activated to R. Therefore, T·I is more stable than T. Allosteric enzymes are also stabilized by binding their substrate, S. Activator

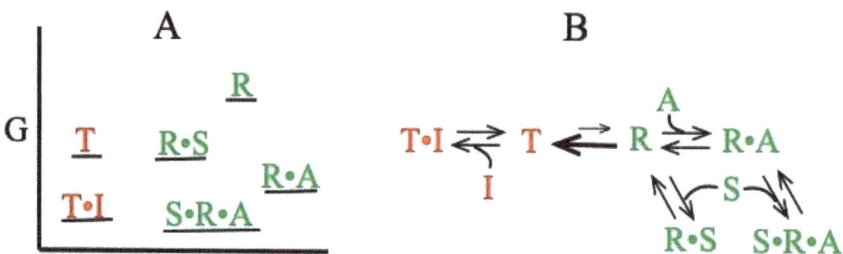

Fig. 8.3 Thermodynamic stability of enzyme conformations. G represents the free energy associated with any molecule. When proteins fold, they reach a stable tertiary structure that reflects their lowest free energy. (**a**) Energy levels; (**b**) Interconversions of T and R, the inactive and active forms of the enzyme. S, substrate; A, activator; I, inhibitor. The bold arrow in B emphasizes that allosteric enzymes will be proportionately more in the T form, since that is the more stable form. Note that ligands always stabilize (lower G) the conformation of the enzyme that binds the ligand

and substrate normally bind at separate sites, so both can bind to and stabilize the R form.

2 There Are Two Types of Allosteric Proteins

At least 30% of all proteins are allosteric in response to some effector. However, allosteric changes have two very different possible outcomes. Most allosteric enzymes, as well as HbA, are designated K-type enzymes or *K*-type proteins because the allosteric effect changes their affinity: the K_M for enzymes or the p50 for HbA (Fig. 8.2). A much smaller fraction of enzymes is designated V-type because the allosteric effect changes their maximum velocity.

This change in V_{max} can be quite dramatic, as exemplified by the enzymes in the blood clotting cascade. Let us first consider how amazing our blood clotting system is, because it is finely balanced between two very different constraints. First, when an injury occurs, it is essential that the wound be closed as quickly as possible to minimize the loss of our very important blood. For such a quick response to be possible, the components for forming the blood clot (Fig. 8.4) must already be available in the blood at all times. And they are.

But second—though all these components must be available—they should never form clots in the absence of an injury. Clotting does occasionally

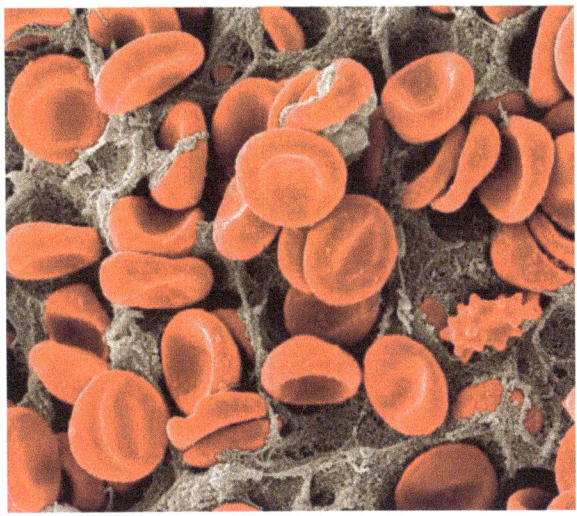

Fig. 8.4 A blood clot: the fibrin mesh is brown, and platelets are red. (Source: Clouds Hill Imaging)

happen, as we get older, and also if we have too much fat in our diet. If such a clot occurs in our limbs, for example, the damage is not that deleterious. Though the tissue just beyond the clot no longer receives oxygen and begins to die, such tissues can normally be regenerated.

But if it occurs in our brain, this is a stroke. It may not be lethal but normally has very detrimental results. If it happens in our heart, this can soon lead to cardiac arrest as heart muscle tissue begins to die.

3   The Nature of *V*-Type Enzymes

Perhaps 99% of all enzymes are designed to have a nearly constant V_{max} or k_{cat}. Therefore, their activity is largely dependent on the available concentrations of substrates. Allosteric *K*-type enzymes may have a conformational change, leading to a shift in their affinity (ΔK_M), but normally have no significant change in k_{cat}. This feature is very logical for metabolic enzymes—those that have some significant catalytic function in central metabolism that must be performed somewhat steadily, even though the actual rate may vary by 2–ten fold, depending on changing physiological circumstances.

However, there are some special situations where an enzyme's activity should normally be at or near zero, yet when a special event requires their activity, they "magically" appear and function at full speed. The solution is to have enzymes that are normally inactive but present at some reasonable concentration, and that can be chemically activated in a very rapid fashion to perform the desired special function.

These are V-type enzymes that demonstrate remarkable changes in their maximum velocity and are found in specialized regulatory systems such as blood clotting and signaling by G-proteins. The clear distinction for such enzymes is that, unlike normal metabolic enzymes, they do not need to be active most of the time.

But the enzymes must always be present, should their activity become immediately needed. Therefore, these enzymes have evolved to exist in two different conformations, one of which is almost inactive, so that when activated, the R conformation can have an enormous relative increase in activity compared to the inactive T conformation (see Fig. 8.3).

Some readers will be aware that in many physiological processes, where an increase in one or more enzymes is required, genes coding for these enzymes can be activated. But this process requires activating the gene with some activator, perhaps a hormone, and then transcribing it into mRNA, and then having the mRNA exit the nucleus to the cytoplasm, where it will be translated into the desired protein. This process requires several hours before the first enzyme molecules begin to appear, and it would be quite unsatisfactory for minimizing the loss of blood after some injury.

Another important feature evident in Fig. 8.5 is the use of a cascade of enzymes to amplify the initial signal of injury into the resulting formation of a clot. There are 13 different proteins that can be involved in forming a blood clot. When they were first discovered, it was not clear what each of these did,

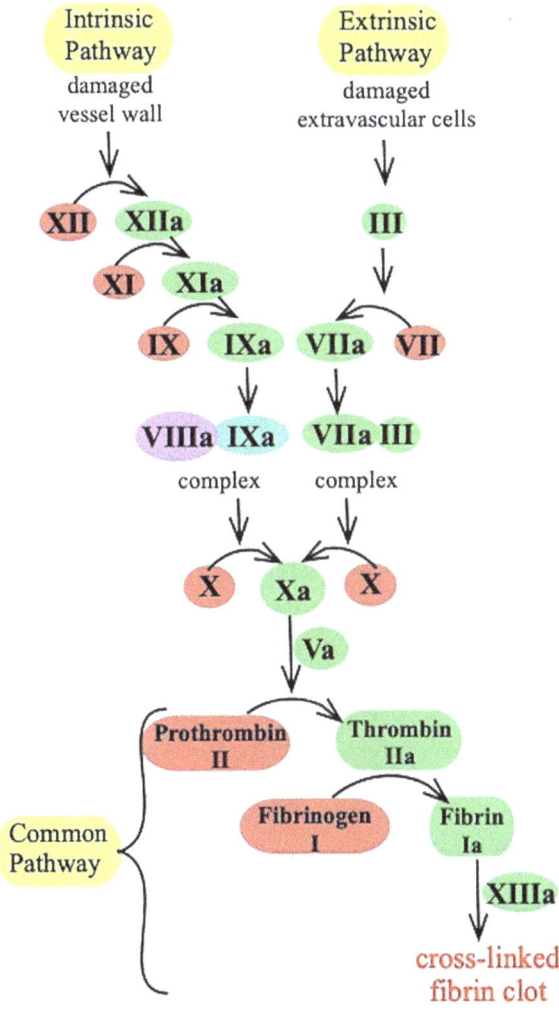

Fig. 8.5 Blood clotting enzyme cascade. Active factors have "a" as part of the name

and so they were simply dubbed "Factor I," etc. (always with a Roman numeral).

We now know that Factor I is actually fibrin, a structural protein that becomes cross-linked to form a mesh, which then attracts and traps platelets to form the flexible plug that can fit into any type of wound and begin to seal it. From Factor V to Factor XII, these are all *proteases*, meaning that their function is to cleave some protein substrate into two separate pieces.

The highest factor, Factor XIII, is not a protease. It is a *transglutaminase*, which catalyzes the formation of a covalent bond between a glutamine residue on one fibrin molecule and a lysine residue on another fibrin molecule, and this secures the stability and integrity of the fibrin mesh that is being rapidly formed.

It was described earlier that hydrolases, such as proteases, are often very unspecific, meaning that they could cleave a substrate protein between almost any two amino acids. This would be true for digestive proteases produced in our intestines, whose function is to take proteins in a meal completely apart, so that the released amino acids can be absorbed by intestinal cells and reused. But for the blood clotting cascade in Fig. 8.5, it is necessary that these proteases be highly specific, and that they cleave only a single specific factor at a very specific site, so as to convert it to the active form.

Each factor normally exists as an inactive precursor, and a part of each of these proteins must be cleaved off by one of the other factors in this cascade to become activated. Two types of nomenclature have evolved to help distinguish these. For most of the factors, the active version, after cleavage by another protease, includes an "a" at the end of the number to indicate that this is the *active* version. Two slightly different naming formats are used for Factors I and II. The precursor, inactive form of fibrin is *fibrinogen*, meaning that it generates the active fibrin. And the precursor form of thrombin is *prothrombin*.

To help understand the need for this cascade, consider Table 8.1, which shows the concentrations of the most important factors. First, remember that

Table 8.1 Key clotting factors

Protein	[protein] (nM)	k_{cat} increase
Factor Xa	170	~10^5
Factor IXa	90	3×10^5
Factor VIIa	10	2×10^6
Thrombin		180
Prothrombin	1400	
Fibrinogen	7600	

these concentrations are throughout the circulatory system, but the clot will be formed in only a limited space close to where the wound is. We see that Factor I, fibrinogen, the precursor to fibrin that forms the mesh, is at the highest concentration. This is logical, as it will take many of these structural fibrins to form the mesh.

Next highest in concentration is prothrombin, the precursor to *thrombin*, which will convert fibrinogen to the active fibrin. But the enzymes needed to activate Factor X and then prothrombin are at much lower concentrations. After all, these are enzymes, and each enzyme molecule can activate many of its target substrate factors every second. Last, notice in Table 8.1 the increase in catalytic activity for each protease upon being activated. Mathematically, this is easy to understand: a normal level of enzyme activity, for the active version, is divided by an extremely low value for the inactive version.

The important fact is that the increase in activity is between 100,000- and two million-fold. This is why blood clotting does not happen normally unless it is activated by the release of calcium or tissue factor at the wound, and then occurs within about a minute.

Humans have multiple disorders involving their red blood cells, of which *hemophilia* is an example. The word literally means "love for blood" and would be an appropriate tag for a vampire. But the word was coined to describe the constant need for fresh blood infusions by hemophiliacs, who have a blood clotting disorder, so that donor blood can remedy this problem on a short-term basis, because blood components have a limited lifetime. The most common problem causing hemophilia is the absence or inadequate concentrations of Factor VII.

Let us summarize *V*-type enzymes in general. Because the precursor version has almost no enzymatic activity, the active enzyme, with a fairly normal level of activity, can now perform some important function. This increase in activity can be quite remarkable. However, as shown in Fig. 8.6, many V-type enzymes also have a meaningful change in K_M. The axes in Fig. 8.6 are on a log scale, meaning that values do not increase 1, 2, 3 …, but increase 1, 10, 100, 1000 … Just examine the horizontal axis.

Figure 8.6 has somewhat unusual units for both axes. This was necessary in order to be able to compare 30 different enzymes for which such detailed measurements were made, since the individual enzymes varied greatly both in their K_M values and in their V_{max} values. Therefore, for each data point, the y-axis shows the ratio of the velocity, shown as k_{cat}, in the presence of a specific effector, divided by the velocity under control conditions with no effector present. That is, on the y-axis, the values show how much faster the enzyme is once it has been activated.

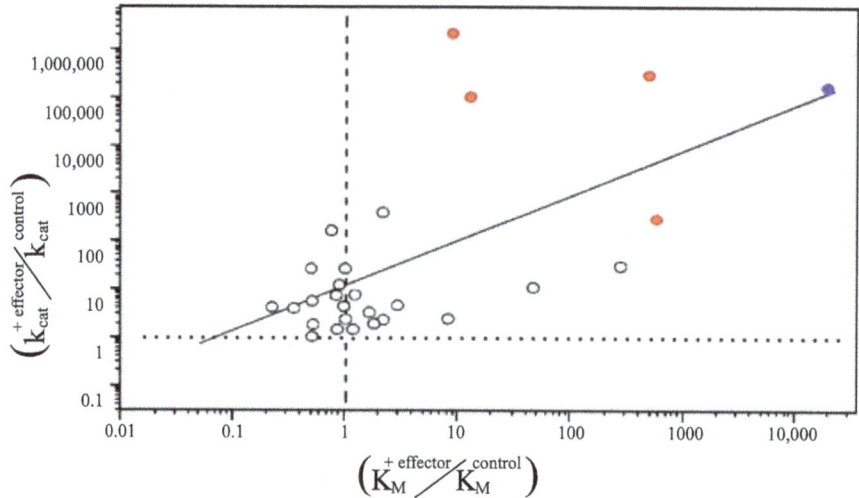

Fig. 8.6 V-type enzymes have dramatic changes in V_{max} but also have some changes in their K_M, like K-type enzymes. For blood clotting enzymes (●) and the G protein (●), the active states are very fast. Note that both axes have log scales

The same procedure was used with the values for the x-axis. Here, we have for each data point the ratio obtained by dividing the K_M value in the presence of the effector by the K_M value for that same enzyme when there is no effector present. For many enzymes in this figure, their K_M ratio stays very near to 1.0, shown by the vertical dashed line, meaning that their K_M changed very little. These enzymes, by chance, were the first ones to be measured in this fashion, which is why they were then called V-type, because only their k_{cat} or V_{max} values increase so much. Note that one of the blood clotting proteases can increase in activity by two million-fold.

A footnote on blood clotting: The amazing process of blood clotting described above is mainly for superficial wounds. These wounds are the most likely to occur to all of us. They are usually not very deep and largely involve bleeding from capillaries under the skin, or veins that are a little deeper. In these vessels, blood flow is not as fast, so that clotting could occur without the fibrin and platelets being swept away from the wound site. This was still helpful because animals and early humans had no bandages, and this minimized blood loss. The slow seeping out of some blood would also have diminished the entry of bacteria. After all, neither animals nor early humans were noted for cleanliness.

But beginning about 5000 years ago, at the dawn of the Bronze Age, humans learned how to make sharp and pointed weapons to be used in battle. These produce deeper wounds, severing arteries, which have higher blood pressure and flow rates. Since then, millions of men have bled to death on battlefields from deeper wounds produced by spears, swords, knives, axes, bullets, etc.

4 G-Proteins

G-proteins are named for binding the nucleotide GTP, which stabilizes the active conformation, and these bind to downstream target proteins, which they then modify, usually by *phosphorylation* (transferring a phosphate group onto another molecule) (Figs. 8.7 and 8.8). As shown in Fig. 8.8, this change in the G-protein is initiated by an external signal (usually a hormone), which binds to a GPCR (G-Protein-Coupled Receptor).

The G-protein *Ras* is normally maintained in an inactive conformation, stabilized by binding GDP very tightly. It requires an activator, GEF (**G**uanine nucleotide **E**xchange **F**actor), to promote the release of the tightly bound GDP, thus permitting the much more abundant GTP in the cytoplasm to bind and stabilize the active conformation. Because the G-protein has a very poor hydrolysis rate, even in the active R form, it takes between 20 and 200 minutes for it to cleave the GTP. This time varies for the many different G-proteins and is therefore the *limited* time period for each such isozyme to remain active. By comparison, the P - proteins (phospho-proteins) that are produced by G-proteins remain in their new conformation indefinitely, unless reversed by a phosphatase.

Thus, G-proteins act as biological time clocks. They have a preset but limited amount of time in the active conformation, stabilized by the binding of GTP. The duration of this active conformation is determined by how slowly the G-protein then cleaves GTP to GDP, which then returns the enzyme back

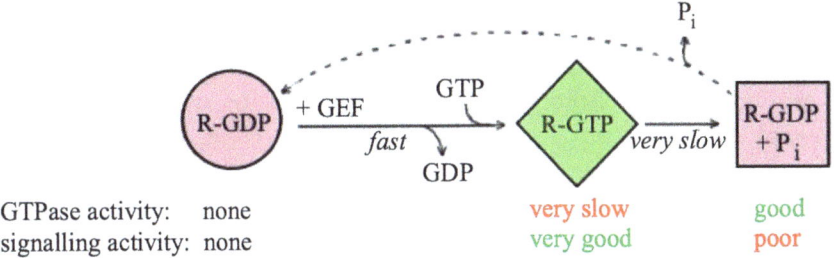

Fig. 8.7 Interconversions of different conformations of *Ras* (R). R-GDP is inactive. The guanine nucleotide exchange factor (GEF) induces the exchange of GDP for GTP to produce R-GTP, the conformation active in signaling

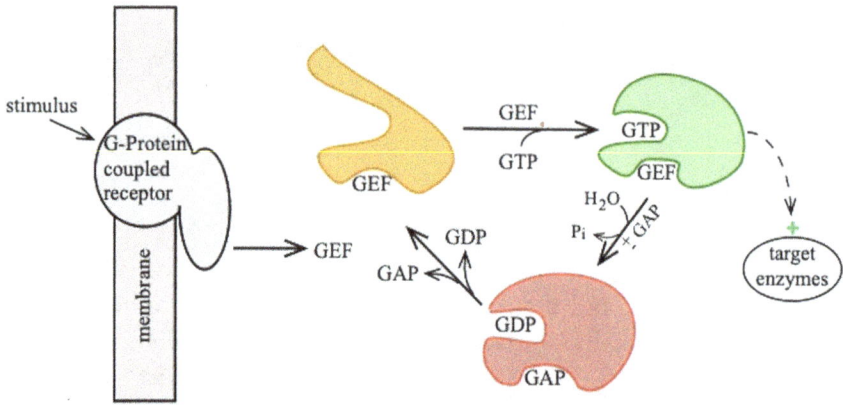

Fig. 8.8 Conformational changes in the G-protein *Ras*. The active conformation (green) requires binding of GTP. After slow hydrolysis of GTP to form GDP, the enzyme changes to very low activity. A guanine nucleotide exchange factor (GEF) promotes the release of tightly bound GDP, so that GTP can again bind and activate the enzyme. GTPase-activating protein (GAP) replaces GEF and speeds up the hydrolysis of GTP and inactivation of the enzyme

to its inactive conformation. Thus, this protein can exist at some reasonable concentration in the cell while having absolutely no catalytic activity. The stimulus produced by an extracellular effector, such as a hormone binding to the GPCR (Fig. 8.9), results in the formation of GEF (Guanine nucleotide Exchange Factor),[1] and this leads to the immediate conversion of the G-protein from the T form to the R form, with an immediate appearance of the catalytic activity at the desired rate.

Resources

Brief description of allosteric enzymes: https://byjus.com/neet/allosteric-enzyme/
Blood clotting cascade: https://www.osmosis.org/answers/coagulation-cascade
Description of G-proteins: https://en.wikipedia.org/wiki/G_protein-coupled_receptor

---

[1] Both GTP (guanosine triphosphate) and GDP (guanosine diphosphate) are guanine nucleotides.

9

Sex and Nucleic Acid Enzymes

Abstract DNA is ideal for information storage because it is a very stable molecule and will therefore be unchanged in a person's descendants, except for occasional mutations. But, because mutations do occur and can sometimes be harmful, the need to have at least two different copies for each gene led to the need for organisms to generally have two sexes, and by mating they contribute these genes to the offspring, so that there would be a greater likelihood that one gene from each such pair would be unchanged and still function normally. Therefore, the process of sex arose simply to ensure this combining of DNAs in any new offspring. Even bacteria exchange DNA. A brief discussion of how DNA was discovered and the history of molecular biology. The process and consequences of DNA replication. To accomplish the complete replication of a human genome requires many origins for replication at the same time. Depiction and description of a mammalian gene with control elements and promoter sequences, TATA box, as well as exons and introns. New mRNAs are protected by a 5′ cap and a 3′ poly-A tail. Description of DNA sequencing methods. Description of DNA fingerprinting using RFLP (recombinant fragment length polymorphism) or snp (single nucleotide polymorphism), or STRs (short tandem repeats). Description of electrophoresis for separating proteins or nucleic acid molecules. Description of DNA repair. Discussion of mutation rates. Long-lived animals have better DNA repair, and larger animals have longer lifespans. The use of methylation to silence genes when they do not need to be active, and enzymes that can reverse that.

Keywords Information storage · DNA · Exchanging DNA · Human DNA composition · DNA replication · DNA helicase · DNA polymerase · RNA polymerase II • RNA primers • Primase • Exonuclease • Promoter • Exon • Intron • Heterogeneous nuclear RNA • mRNA • 5′ cap • Poly-A tail • TATA box • RFLP • snp • Electrophoresis • STR • Methylation of CpG • Transcription factors

What is DNA good for? Why do we need it? This "blueprint" for making any particular organism, such as you, the reader, requires a set of specific instructions for making the many different proteins (the worker bees) that know how to synthesize all the molecules needed to make every type of cell and tissue in our bodies. Ideally, these blueprints should be very stable, so that, once a good working set has been achieved, they will remain stable and continue in later generations, largely unchanged. That is why your children resemble you.

DNA is ideal for this purpose because DNA is unusually stable: it takes almost one million years for a DNA molecule to be hydrolyzed (Greek for splitting by water) (Fig. 3.3). However, changes to DNA (mutations) can and do occur because of radiation (see Sect. 10). And that is why sex became necessary for life to continue. Readers may have various views about why sex is necessary, from RomCom novels to spicy movies. But sex first became established at the beginning of life with bacteria because, in the biological sense, sex simply involves an exchange of genes, or an exchange of DNA.

Let us consider a simple *E. coli* bacterial cell, whose ancestors were the first cells to exist. Any individual *E. coli* cell will experience mutations to its DNA—this is inevitable (Sect. 10). But, it very early became a feature of such cells that they would occasionally join side-by-side so that their membranes could fuse, which enabled molecules to move between the two bacterial cells, and so exchange some of their DNA. This would have been the first occurrence of sex—the swapping of DNA. Because, when a cell that had experienced a mutation that normally might kill it was able to swap DNA with another cell where that section of DNA had remained normal, the "damaged" bacterial cell now had a good copy of that DNA and could survive.

Such exchanges of DNA were random in terms of what sections of DNA were exchanged, but this still allowed some cells to survive and then make more copies of themselves when the included new DNA had the needed correct segment.

For more complex organisms, this is accomplished by mating, where one set of the chromosomes comes from each parent, so that the fertilized egg now has a duplicate copy for almost all genes. Even if one gene from one parent has become mutated, the one from the other is usually good.

1 The History of Molecular Biology

The definition for the field of molecular biology has an interesting history. In 1953, James Watson and Francis Crick, after months of making possible three-dimensional structures of a DNA double helix, finally achieved a successful model and published their results. Their efforts were initiated by some results with the crystal structure of DNA by Rosalind Franklin. Franklin's crystal structure did not have very high resolution, meaning that the details of the DNA double helix were somewhat blurry and unclear, but it was just good enough to suggest that DNA existed as two strands, somehow bonded together.[1]

We have already seen several types of weak bonds in protein structures (Table 6.4), and these are the same types of bonds that might help to stabilize two DNA strands in a complex (Figs. 9.1 and 9.2).

At the time that Crick, an Englishman, and Watson, an American, first encountered each other, Crick was 37 years old and still a graduate student working for a doctorate in physics. Watson was 12 years younger and had already completed a doctoral degree in zoology. Watson had then traveled to

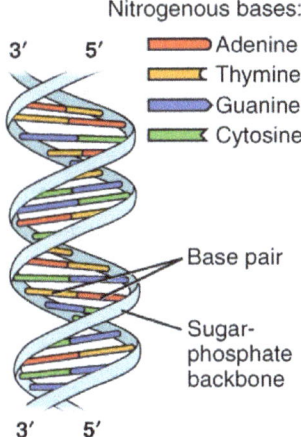

Fig. 9.1 DNA double helix. The two separate DNA strands are held together by base pairing and must be antiparallel. Each strand is complementary to the other strand

---

[1] Many people have claimed that Rosalind Franklin should have shared the Nobel Prize with Crick and Watson and that she was cheated. The reality is that Franklin died from cancer in 1958, and the prize was not awarded until 1962. When the prize was first established in 1903, it was given to younger, living scientists to help them with their research. Many years later, it evolved into an award recognizing older scientists for some remarkable achievement.

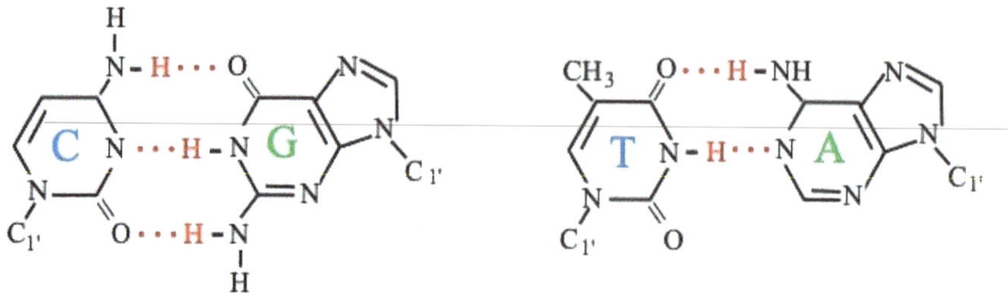

Fig. 9.2 Base pairing. In the DNA of all species, %G = %C and %A = %T. Only the bases are shown, and the rest of the nucleotide is attached via the $C_{1\text{ (carbon 1)}}$ of the pentose. $C \equiv G$ has 3 H-bonds, $A = T$ has 2 H-bonds. Because one of the bases above is thymine, this would represent a base pair in DNA. If thymine is replaced by uracil, it could then be a base pair between DNA and RNA during transcription

Cambridge, England, for some additional postdoctoral training. It is an interesting observation that the financial fellowships that these two men had each obtained were not intended for them to work on DNA. They were absent from their intended work while pursuing what turned out to be more significant results.

With much more experience in atomic-level structures, Crick led the two in constructing physical models to simulate what a double-stranded DNA complex might look like. Hydrogen bonding was a highly likely choice, and as shown in Fig. 9.2, a base pair is always formed between one pyrimidine and one purine base, and this was actually first proposed by Crick and Watson. Then, even though these two types of bases are different in size, when paired in this fashion, all such pairs will have the same general width between the two DNA strands. Also, in order that the bases would be able to form H-bonds between the two strands, the DNA backbone had to be formed by alternating ribose and phosphate groups (Fig. 9.3). This original paper was hugely important because it demonstrated that the two DNA strands could be held together by hydrogen bonds. The two strands also had to be antiparallel, as shown in Fig. 3.2 and in Fig. 9.1.

What made this original model for the DNA structure so remarkable was that it gave a clear depiction of the DNA at the level of the atoms that formed it. In 1959, a new journal was introduced to emphasize molecular structures, named *The Journal of Molecular Biology*. Then, in 1965, Watson published a detailed description of what was then known about DNA, titled *The Molecular Biology of the Gene*. This book was so successful that the term molecular biology, originally intended to refer to molecular-level definitions of all biological compounds, was now co-opted to refer to any research with DNA, and then RNA.

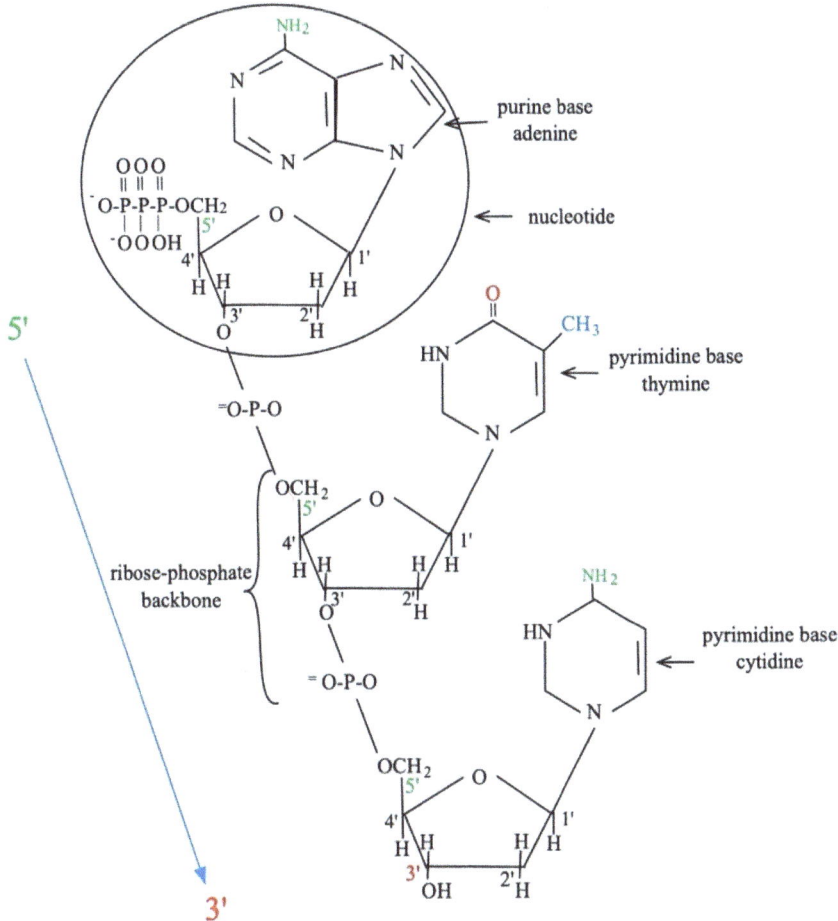

Fig. 9.3 A model of a DNA trinucleotide. To enable base pairing, the bases cannot be part of the backbone but must point inward toward the complementary strand

Let us first consider how our genome is organized. It may be a surprise to many readers that only a very small portion of our genome actually contains genes that code for proteins or RNAs (Fig. 9.4). Over millions of years of evolutionary time, our genome has acquired a lot of apparently nonfunctional or useless sequences. Many of these come from infections with bacteria or viruses, when their genetic material was accidentally inserted into an existing chromosome in a reproductive cell: sperm or ova. Almost always, these accidental additions remain silent—they are not expressed into proteins. Therefore, it was possible to tolerate their inclusion.

It is interesting that we never evolved processes for maintaining our genome as mostly coding DNA. By comparison, *E. coli* DNA has almost no introns or

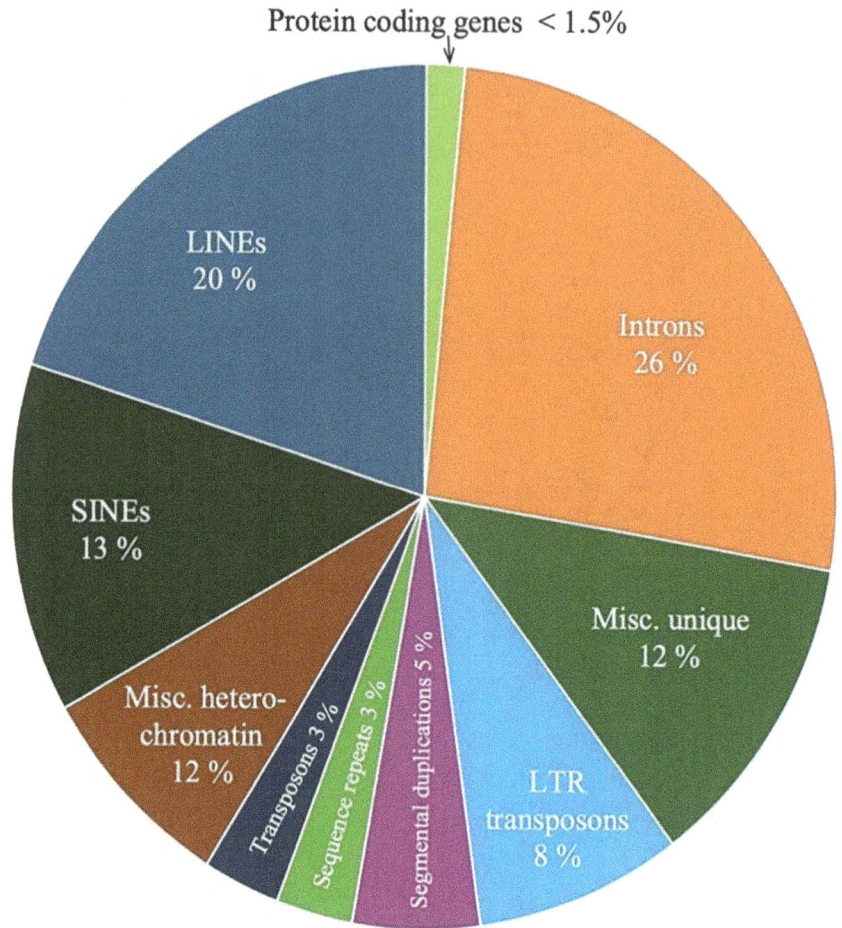

Fig. 9.4 Composition of human DNA

transposons and is more than 90% coding DNA. It takes a lot of ATP to synthesize the deoxynucleotides in DNA and then also to join them into the final DNA double helix. It is likely that a reasonable amount of what was earlier described as "junk DNA" serves a function yet to be discovered. We must also be aware that gene totals (Table 9.1) are not absolutely correct. With continued sequencing for each species, the totals continue to be modified slightly.

While eukaryotes have more genes than simple bacteria, there is not a good correlation between the total number of genes and how advanced or complex an organism may be. It is also possible that DNA serves functions other than coding for proteins. Because it is energetically costly to synthesize DNA, it is unlikely that this would occur or be maintained if it did not serve a beneficial

Table 9.1 Number of genes for different species

Species	Number of genes
Mycoplasma genitalium (bacteria)	465
Streptococcus pneumoniae (bacteria)	2236
Escherichia coli (bacteria)	4377
Saccharomyces cerevisiae (yeast)	5770
Drosophila melanogaster (fruit fly)	13,379
Caenorhabditis elegans (worm)	19,427
Homo sapiens	22,687
Arabidopsis thaliana (weed)	28,000
Mus musculus (mouse)	29,000
Oryza sativa (rice)	37,544

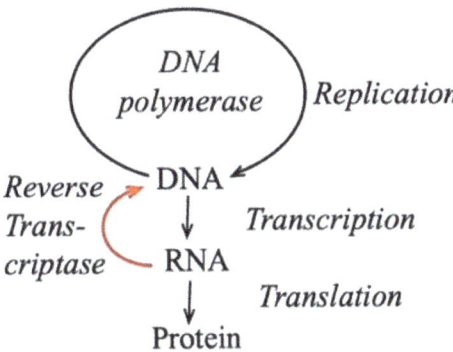

Fig. 9.5 The central dogma of molecular biology

function. It is interesting to see that our last universal common ancestor, LUCA (see Fig. 5.9), has been estimated to contain about 2600 genes, a number very similar to some simple bacteria existing today.

Before exploring how DNA is synthesized, let us consider how the main processes involving DNA all work. The different steps in Fig. 9.5 are so well established that this is now referred to as the *Central Dogma*. An important aspect is that this is almost always a one-way process. An exception was later discovered with *reverse transcriptase* enzymes in bacteria and viruses (see Chap. 12).

2 Replication of DNA

Let us now see how DNA is duplicated, a process occurring routinely both in laboratories and in all cells that continue to replicate, and (see Fig. 9.6) a source of genetic flaws when it does not proceed correctly. This figure briefly

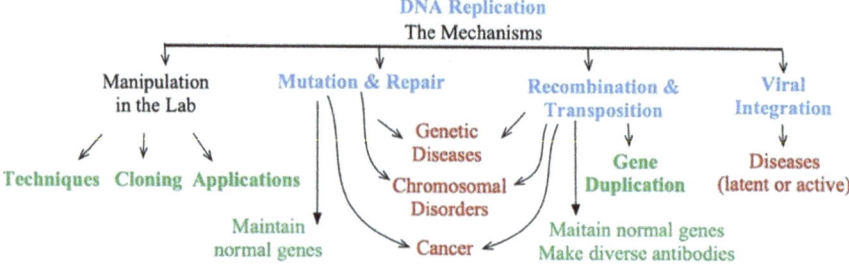

Fig. 9.6 Possible outcomes from replication of DNA

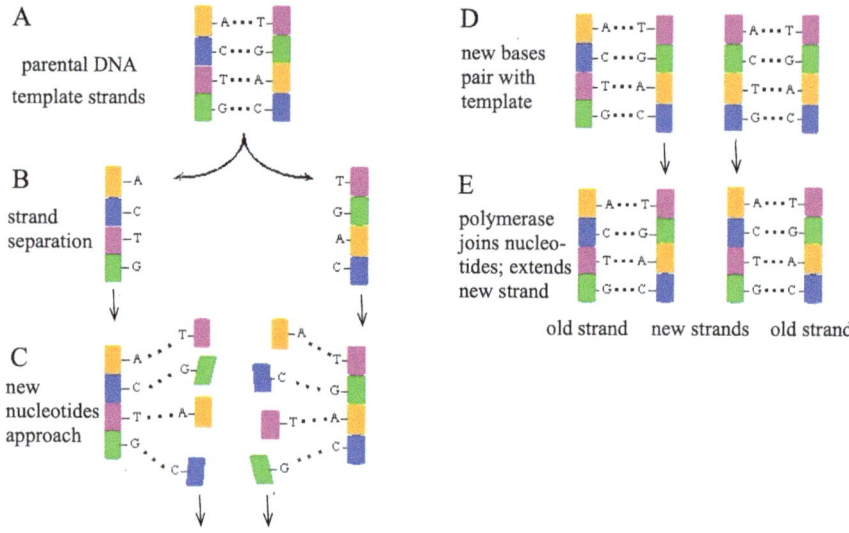

Fig. 9.7 Replication of DNA. A, the base-paired parental strands. B, two strands are separated by *helicase* (not shown); C and D, appropriate nucleotides with the correct base for bonding align along the template strands. E, DNA polymerase then joins the individual nucleotides to make two new strands, each of which base pairs with a parental strand

depicts mechanisms for duplicating DNA and possible associated outcomes that are good or bad.

To begin replication, the two base-pairing strands must be separated (Fig. 9.7), and then each one can serve as a template to form bonds with the appropriate individual nucleotides, having the base that will correspond to the template strand. As new nucleotides become appropriately aligned via base-pairing, DNA polymerase slides along and catalyzes the formation of a bond between adjacent nucleotides (Fig. 9.7e).

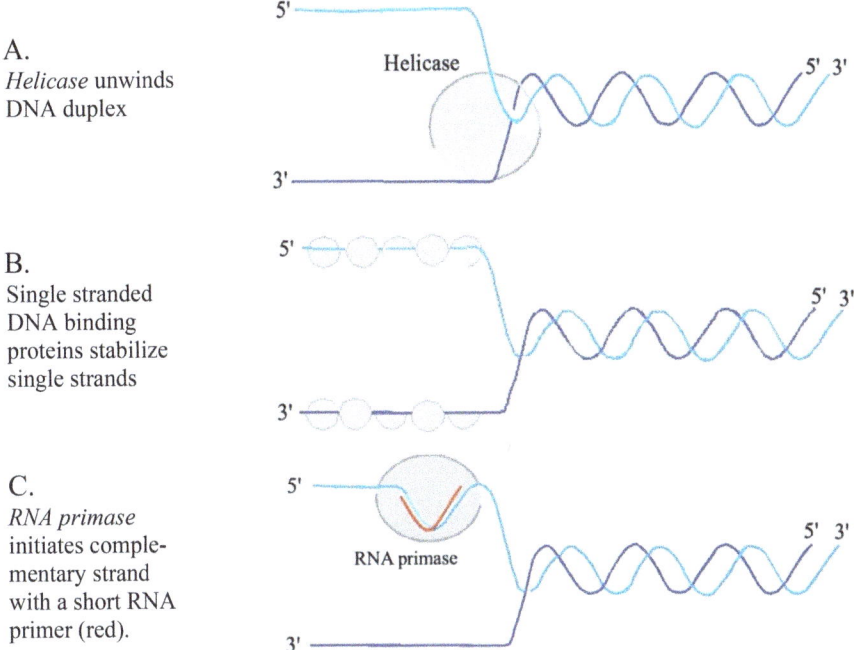

Fig. 9.8 The initial steps in DNA replication

Mammalian DNA polymerase is unable to bind with a single first deoxynucleotide and then add a second deoxynucleotide to it so as to begin extending a new DNA strand. It can only extend an existing strand, even if it is short. This problem is solved by having a short starter nucleic acid, called a *primer*, which is actually composed of RNA nucleotides, and synthesized by the enzyme *primase*. In Figure 9.8C, the RNA primer is red. Such a primer may be as short as 4 nucleotides in laboratory experiments but is generally a little longer. Once the newly duplicated DNA strand is complete, the small initial RNA segment is removed by an *exonuclease*. This is a separate activity of the DNA polymerase. This short gap is then replaced with deoxynucleotides by a DNA repair enzyme, thereby making a complete DNA strand.

E. coli and humans were the two species for whom DNA replication was first studied. In bacteria, the polymerase had a rate of joining at about 252 bp/minute. Simple math showed that for a total genome of 4,200,000 bp, it would take a bacterial cell, going through cell division, 278 h just to duplicate the cell's DNA (Table 9.2). But the observed time for them to do this in a laboratory dish was about 20 min.

Table 9.2 How many origins for replication?

Phylum	Species	Genome (bp)	Replication time (1 origin)	(1 origin/ chromosome)	Actual	# of origins?
Bacterium	E. coli	4.2 × 10⁶	278 h	278 h	0.3 h	~900
Mammal	H. sapiens	3.3 × 10⁹	~900,000 h (~103 yrs)	~40,000 h (~4.6 yrs)	20 h	~2000

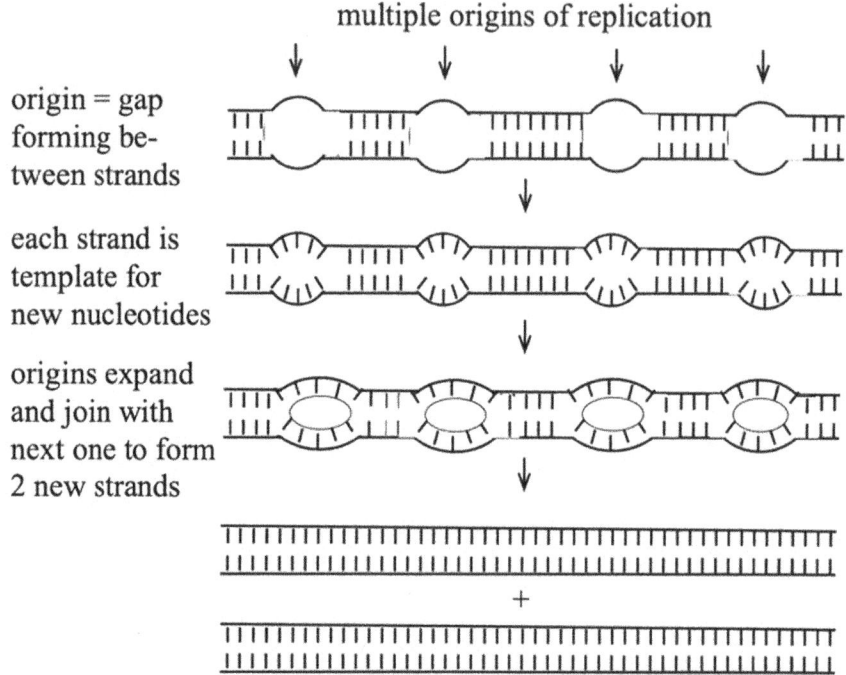

Fig. 9.9 Expanding origins of replication. As the origins expand in both directions and unite with neighboring origins, two new DNA strands will be completed, leading to a duplication of the original DNA

3   Many Origins of Replication Are Required for Timely Duplication

Clearly, DNA being completely duplicated in a reasonable time could occur only if there were many origins, so that multiple *DNA polymerases* could work at each location/origin simultaneously and complete the replication more quickly, so that the bacterial cell would not remain in this compromised position for too long. This led to the theory or model presented in Fig. 9.9.

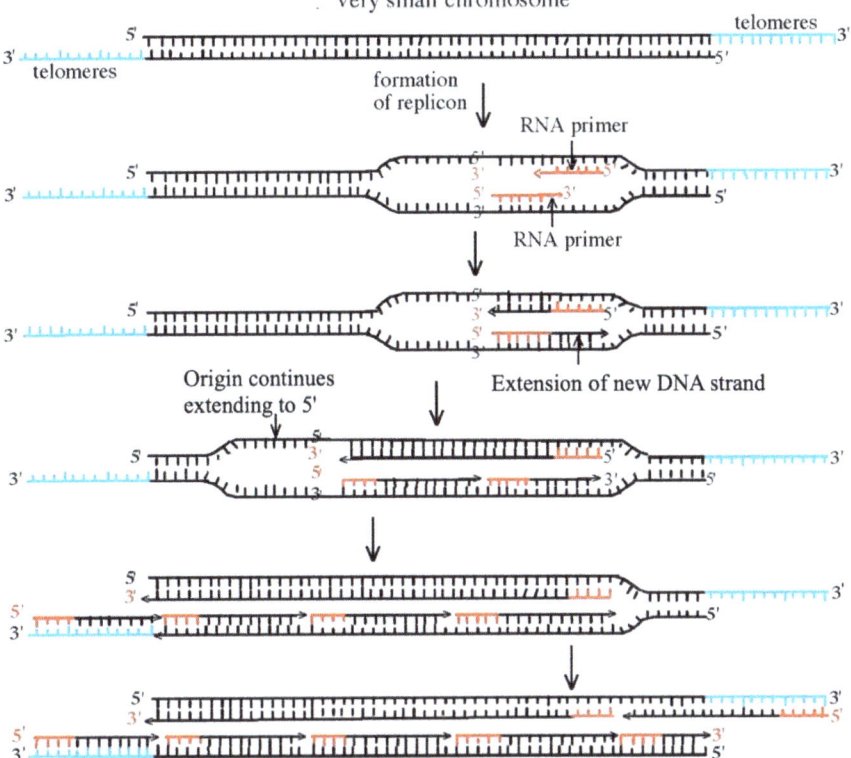

Fig. 9.10 Multiple origins on the lagging strand, shown by red primers. Note: for simplicity, the origin of the replication bubble at the top is shown extending only to the left. In reality, it would extend equally in both directions. The blue extensions are telomeres

The widening of such an origin is illustrated in more detail in Fig. 9.10. Note that when the two strands separate, the new strand being synthesized to the upper, 5′ strand must start at the far right because it must be antiparallel. As the origin continues to expand toward the 5′ direction, the new strand simply continues to be duplicated in the same direction. But for the lower strand being duplicated, the new strand must go in the direction opposite from how the origin is widening and must be antiparallel. As the origin continues to expand toward the 5′ direction, the new strand simply continues to be duplicated in the same direction.

This is made possible by the repeated use of primers as the new template strand becomes accessible toward the 5′ end, where new primers can bind, which the polymerase then extends in the 3′ direction. DNA polymerase then removes the primers, with an *exonuclease* activity (cutting DNA molecules from an end), and replaces them with the correct deoxynucleotides. These smaller segments will then be joined by a *DNA ligase* (from the Latin *ligare* = join).

The use of such primers is also very important in laboratory experiments. If some portion of a gene's sequence is known, this sequence can be duplicated by scientists by chemically synthesizing a primer, and then applying it in a laboratory experiment with some DNA sample. Because this artificial primer is complementary to a portion of some gene in such a sample, it will base pair there, and with the addition of DNA polymerase plus all the deoxynucleotides, an artificial duplicate of the DNA sequence can be made. Even if this does not produce the complete DNA sequence of this gene, the process can now be repeated, using a primer for the complementary strand of that DNA, proceeding in the opposite direction, and this would then continue toward the beginning of the gene being duplicated.

4 Overview of Molecular Biology

The field of molecular biology has become so vast and complex that here we will examine the most important components and processes, which are diagrammed in Fig. 9.11. The diagram features a single gene and emphasizes key features. At the top, we see the most common way that double-stranded DNA molecules are presented. The upper strand always has the 5'-end at the left. This is the beginning, or first nucleotide, in this DNA strand. The lower strand goes in the opposite, or anti-parallel direction, so that it begins at the far right.

exon: expressed sequences containing codons for the protein,
intron: intervening sequences with no information for the protein.

In this simplified diagram, regulatory sequences that control whether the gene will actually be transcribed are to the left side, and the sequences coding for the actual RNA that will be transcribed are to the right. This should seem logical because, while each cell has every gene contained in our 23 chromosomes, all genes should not be expressed in all cells. We don't want our brain cells to make muscle proteins.

When a gene codes for a protein, only a limited few segments along the gene will actually contain information for the amino acids included in that protein (Fig. 9.11).

These are the exons, for "expressed + on." Between the exons are the much larger introns in gray, for "intervening + on," which are included in the transcribed *hnRNA* but are removed during splicing. Exons are shown in dark

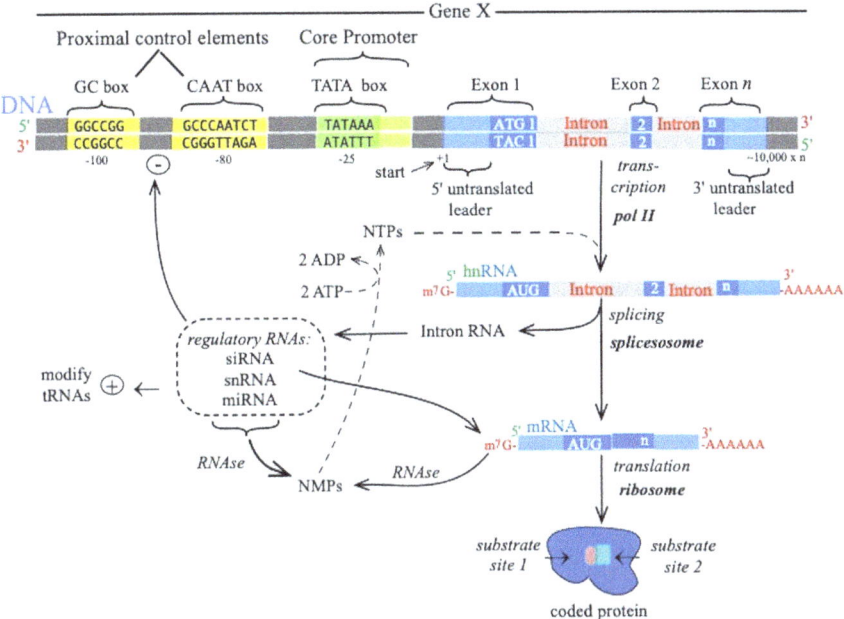

Fig. 9.11 Regulation of gene expression. A single gene contains n exons and n-1 introns. Internal exons contain only coding sequences; the 5′ and 3′ exons also contain untranslated or noncoding sequences. Transcription of the gene starts after the TATA box and produces a direct transcript of the DNA template, as an hnRNA (heterogeneous nuclear RNA) transcript. This initial transcript still includes all intron sequences. Internal introns are removed by splicing to yield the mRNA. Upstream of the *Start* site, the gene includes noncoding sequences at which regulatory proteins bind. Numbering of the nucleotides begins with +1 corresponding to the first nucleotide in the hnRNA. Sequences preceding this position have negative numbering. Because an average exon contains about 150 nucleotides, and DNA is about 1.4% coding, a gene with n exons contains about 10,000 × n nucleotides. Regulatory RNAs can block transcription and translation or promote the maturation of tRNA. It is not definite how regulatory RNAs are processed. *NMPs* nucleoside monophosphates (nucleotides)

blue. The initial and terminal exons also contain some noncoding sections, shown in light blue. This is because transcription begins at a sequence that the polymerase recognizes, and translation always begins at the AUG start codon, which are at separate positions. There is also a STOP codon at the end of the coding hnRNA, but the transcript normally extends beyond this STOP codon.

Going downward from the exon region, we see the initial transcript, formed by *RNA polymerase II*, which is a direct and complete copy of this gene, starting at nucleotide 1, and therefore containing the sequences for the noncoding regions (light blue) plus the introns (gray) as well. This larger initial transcript was discovered much later than mRNA and is labeled hnRNA, for *heterogeneous nuclear RNA*. Unlike the gene, which contains deoxyribonucleotides, the hnRNA contains only ribonucleotides. And it is still in the nucleus and is "heterogeneous" because the sections copied from the introns have not yet been removed.

Cells have many *ribonucleases*, enzymes that cleave or hydrolyze RNA, commonly termed *RNAses*. To protect this newly synthesized hnRNA from *RNAses*, protective elements have been added during or after the synthesis process. Because the 5′ portion of this new hnRNA is made first, and it may take a few minutes to synthesize the entire molecule, the 5′ end is protected from *5′-exonucleases* while the full hnRNA continues to be synthesized, by the enzymatic addition of a cap: 7-methyl-G, which is not recognized by *5′-nucleases* and thus protects this end of the new hnRNA from being degraded. When the new hnRNA is completely finished, a special *polymerase* adds several hundred AMPs to the last nucleotide. This is designated as a *poly-A tail*, shown in red.

There are often more than 100 A's in this extended tail. The *3′-exonucleases*, so named because they degrade RNAs from this end, have no difficulty cleaving the AMPs off one at a time. But to chew through a tail with more than one hundred A's takes many minutes at least, and this therefore defines the lifetime for the new mRNA to be useful for translation. This is a remarkable system for defining a limited time period for a new mRNA to function in making the coded new proteins. Should the cell or the tissue need more of this particular protein, regulatory signals will facilitate additional transcription of the specified gene.

Readers may see the above process as wasteful or energetically expensive. But it should become clearer that for higher eukaryotes (from the Greek: *eu* = good; + *karyon* = kernel, i.e. the nucleus), especially mammals, control of these processes is most important. And because animals are generally good at finding food, thereby providing energy for these various biochemical processes, such animals can afford to use more ATP simply to assure fairly precise timing or balance in synthesizing RNAs and proteins. A lot of a cell's energy is used simply to assure this type of control.

At this point the new hnRNA is still within the nucleus. It must now be converted into the desired messenger RNA (mRNA) by removing all the sequence portions corresponding to introns in the gene. An enzyme complex, the *spliceosome*, then slides along this newly made hnRNA and recognizes the sequences that define *exon/intron boundaries*. It first cleaves the new hnRNA before an intron, and again just after the intron. The excised intron can then diffuse away, and the *spliceosome* joins the 3′ end of the previous exon to the 5′ end of the following exon. This produces the new mRNA, containing the continuous message for the amino acids that will form the new protein the gene codes for.

5   Other Control Regions in a Gene

If we now examine the left side of the double-stranded DNA (at the top of Fig. 9.11), we see some short sequences highlighted in color. When these sequences were first observed to have control functions, they were frequently shown in a box, or highlighted as in Fig. 9.10, and so the term "box" became quickly associated with these control elements. Because these sequences occur before the start nucleotide for transcription, they have negative numbers to indicate their distance from this start site.

The TATA box, in green, is where the polymerase attaches to the DNA before sliding along to the ATG codon. Thymine (T) occurs only in DNA, and is replaced by uracil (U) in RNA. Therefore, the ATG start codon in DNA becomes the AUG start codon in mRNA. However, to help the polymerase get here somewhat more quickly, additional sequences upstream are recognized by additional control proteins. With nucleic acids, the term *upstream* always means more to the 5′ end, or to the left, and *downstream* means more toward the 3′ end, or to the right. Like a raft in a river, the *polymerase* always moves downstream.

The benefit of these additional control boxes, in yellow, is that novel control proteins, sometimes produced by special physiological conditions, for which a new protein is suddenly needed, can bind here and increase the frequency of the polymerase landing at the TATA box. While such diagrams are always linear and therefore make the boxes that are further upstream appear to be too remote for any control protein binding there to still be able to connect with the polymerase at the TATA box, we should be aware that DNA is not rigid like a wire but loose and floppy like a cooked spaghetti noodle. It can therefore bend and twist within the nucleus, and sometimes an upstream box will be much closer to the TATA box, so that control proteins can assist the polymerase in binding.

6   Processing RNA Transcripts

In the center and left of Fig. 9.11, several other RNAs are shown, and their formation and use are expanded in Fig. 9.12. Here, the primary transcript is the same as the hnRNA in Fig. 9.11. It is then spliced as described above to remove introns. The nomenclature can become confusing. For example, ncRNAs can include all the other extra RNAs.

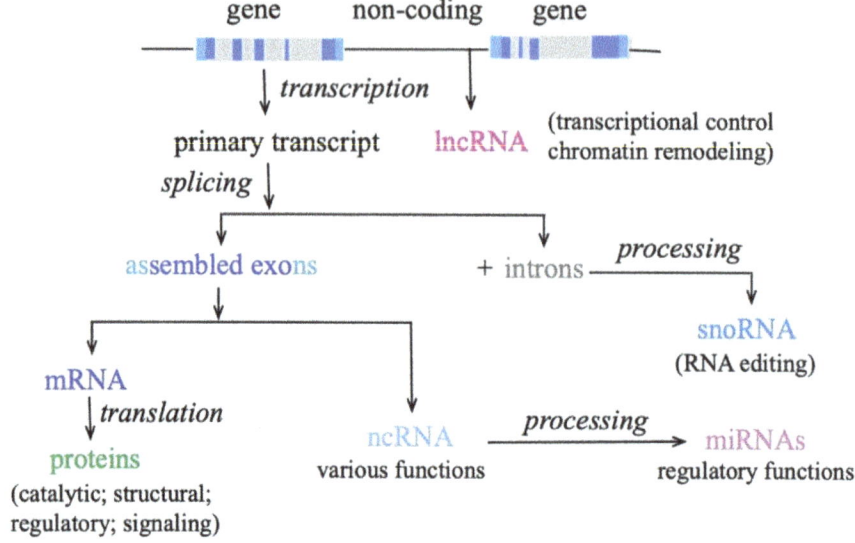

Fig. 9.12 Processing the primary RNA transcript. The processing of the transcript removes introns and leads to the coded protein, plus multiple different RNAs that have various regulatory functions. lncRNA = long noncoding RNA; snoRNA = small nucleolar RNA; ncRNAs include rRNAs, tRNAs; miRNAs = microRNAs

These RNAs are normally fairly small, and there are almost 100,000 such RNAs. Because these have important regulatory functions, many scientists consider them to be genes, thereby expanding the number of human genes to much greater than 100,000.

Before continuing, let us summarize the overall processes, as shown in Fig. 9.13. At the top, replication of DNA by *DNA polymerase* duplicates both strands to provide a new pair with one sense and one template strand. Note that the template strand is complementary to the sense strand and is synthesized in the opposite direction. This new template strand can be used by *RNA polymerase* (enzyme transcribing DNA to RNA) to make an mRNA copy that has the same sequence as the original sense strand. Note that the new mRNA has all T's replaced by U's, so that AUG is the initial codon. This figure has been simplified to omit depicting the hnRNA, which is the initial product made by RNA polymerase. Here, we go directly to the mRNA so that we can also see how the ribosome then slides along this template to translate it into protein.

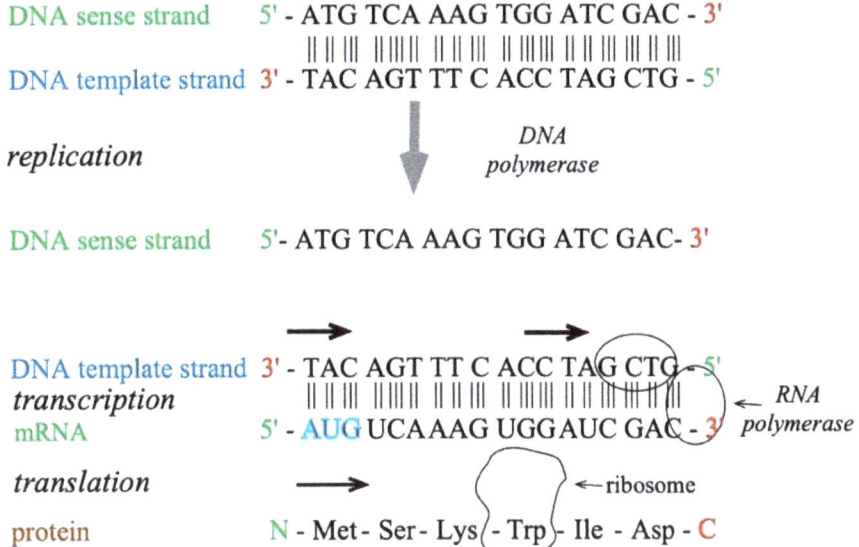

Fig. 9.13 Summary of transcription and translation. RNA polymerase slides along the sense strand to make the mRNA copy, which now starts with AUG, and this is translated to make the specified protein

7 Improvements in DNA Sequencing

The chemical procedures necessary for obtaining the sequence of even a short strand of DNA were very laborious and difficult until the 1970s. Each decade thereafter saw remarkable new advances in technology, instrumentation, and the use of specific enzymes from microbes, such as the *restriction enzymes* briefly mentioned in Chap. 4. It took more than ten years to achieve the first sequencing of the human genome at about 90% accuracy. And this was accomplished by the combined effort of many laboratories, each of which focused on only a limited and specific part of some chromosomes.

Today, the sequencing of DNA has become so automated that the human genome is known to an accuracy greater than 99%, and we also know the genomes of over a thousand different species. Although the entire sequence of the human genome has now been sequenced thousands of times, our knowledge or confidence as to which DNA sequences are genes is not yet completely certain. See the discussion about possible junk DNA sequences above.

Because of this increased speed and economy in DNA sequencing, many readers are familiar with the expectation that a trace of DNA at a crime scene can be sequenced and then compared with similar sequences from any

Table 9.3 Costs for sequencing a full human genome

Time	Methodology	$/genome
1980–2000	Sanger (chemical)	>$100 M
2005–2009	Shotgun/microarray	>$50,000
2010	Nano-array	~$10,000
2015	Next generation	~$1000

potential suspects. The best samples, such as skin or hair, would contain cells, and DNA can then be extracted from these and sequenced.

The cost of sequencing DNA has declined enormously (Table 9.3), going from approximately $100 million for the original human genome to about $1000 for a comparable effort today. But that is still too expensive for most routine purposes. Because most police or criminal laboratories simply do not have the budget for the equipment and chemicals needed to do that routinely with any individual crime. Therefore, other, simpler procedures are used for such purposes.

8 The Importance of Electrophoresis

In order to study DNA strands or individual proteins, they must be separated from other such molecules in a sample mixture. Since both proteins and DNA segments vary in their size, and therefore their total charge, these features are used to separate or isolate them. There are several different technical approaches for this, and one of the most frequently used is electrophoresis (Fig. 9.14). Electrophoresis (from Greek, meaning "to be carried by electricity") is a technique that is fairly inexpensive and widely used.

It requires a simple apparatus (Fig. 9.14 shows an example), consisting of a supporting stand that holds two parallel glass plates. These are held about one-fourth of an inch apart by spacers at the bottom and top, and a slurry of polyacrylamide is gently poured between the plates. This acrylamide will cool and become a little more firm, somewhat like jello.

The acrylamide molecules assume a shape somewhat like very small marbles, which stack but have small spaces between them through which protein or DNA molecules can slowly flow. They move because the unit contains a buffer, assuring that the molecules to be separated have a negative charge (see Fig. 9.1). The apparatus has electrical connections, and the anode—it attracts anions (negatively charged)—is at the bottom.

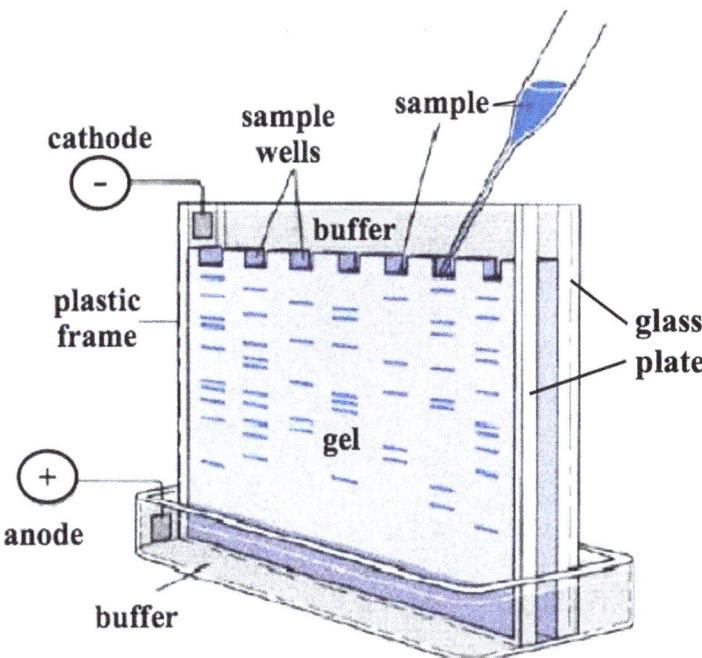

Fig. 9.14 Electrophoresis: samples are applied in slots at the top of the gel and then separated as they move along parallel vertical lanes due to the applied current. Each vertical lane contains a different starting sample. Protein bands are visualized by staining with the dye Coomassie, while DNA binds the dye Ethidium

When the current is turned on, the sample molecules become attracted to the anode and begin to flow downward. The smaller molecules can more easily slip between the acrylamide beads and reach the bottom first. Larger molecules will be more retarded and remain nearer the top.

The apparatus depicted is vertical, and the acrylamide matrix is not easy to manipulate. Therefore, the glass plate unit is removed and placed in a shallow pan. The upper glass plate can then be gently lifted away. A square piece of nitrocellulose paper is then laid gently over the acrylamide gel for about an hour, and this permits the movement, by osmosis, of enough molecules in each of the bands onto this paper. The paper can then be immersed in a solution that will stain the type of molecule adsorbed, and so make these bands visible.

Some DNA molecules can be quite large, so that agarose is used as the separation medium, and because agarose is looser, the apparatus is in a horizontal position.

9 Procedure for DNA Fingerprinting

Many readers may have seen crime shows on television where *DNA fingerprinting* is frequently used to prove a suspect guilty of a crime. This is not done by DNA sequencing. What is actually used is a much simpler and also more limited examination of DNAs from whatever samples are available. A very common procedure uses a method called *RFLP* (restriction fragment length polymorphism). The same procedure is used by companies that examine one's ancestry from a simple sample of saliva.

What is a *polymorphism*? A polymorphism is a single position in the genome sequence for which two or more alternative alleles are present at an appreciable frequency, usually at least 1% of the human population. 1.42 million polymorphic sites have been identified in the human genome. This is a very low number. Our genome contains 3.1 billion nucleotides, so less than one in 2000 has been mutated or altered in many people.

If only a single nucleotide is altered, this is defined as a *snp* (**s**ingle **n**ucleotide **p**olymorphism; Fig. 9.15). But often, two or more nucleotides in sequence may be altered. Restriction enzymes are very exact as to what sequence they will align with (Fig. 3.2), so if only a single base/nucleotide in such a short sequence is altered, the restriction enzyme will no longer bind and cut there. Alternatively, a new site may be formed where it had not existed. This alters the size of the DNA fragments that might be obtained (Fig. 9.16). What is observed in a crime lab or ancestry lab is the pattern of these fragments, separated by size with electrophoresis (Fig. 9.17). Can readers determine if either suspect appears to be guilty?

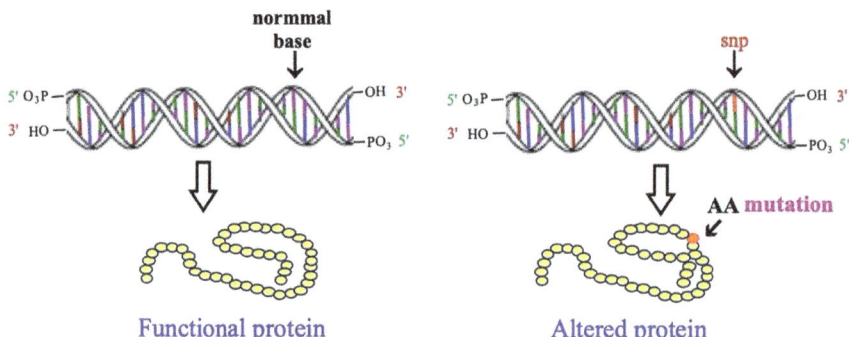

Fig. 9.15 A single nucleotide polymorphism (snp). A single nucleotide mutation can change a codon, producing a different amino acid (AA) in the final protein

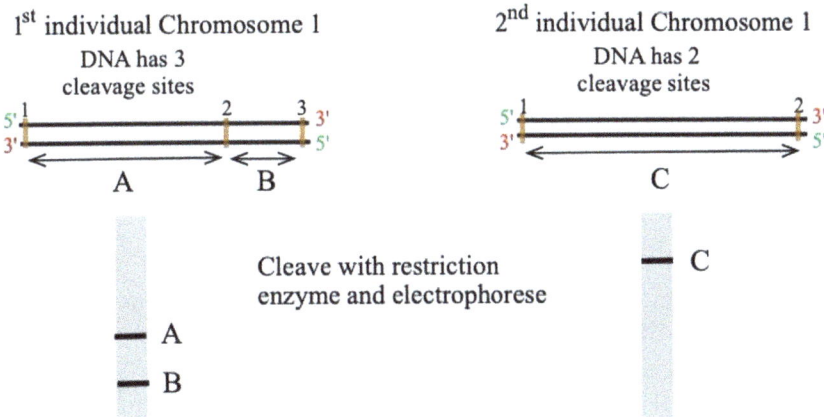

Fig. 9.16 DNA polymorphisms: a single base difference between homologous DNAs from different individuals causes a restriction site to be added or lost

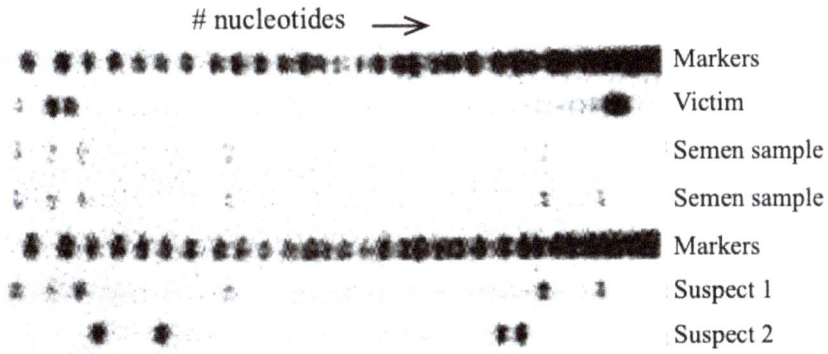

Fig. 9.17 The use of RFLP to compare fragment patterns from a crime sample with two possible suspects. What is shown here is the electrophoresis pattern where migration is from right to left, with the largest DNA fragments at right and the smallest at left. As shown in Fig. 9.11, this electrophoresis was actually done in the vertical direction. DNA samples separate in the horizontal direction. (Image source: Knight, Pamela. "Biosleuthing with DNA Identification." *Bio/Technology*, vol. 8, no. 6, 1990)

This same technique is also useful in ancestry studies. Polymorphisms, or snps, are inherited. Therefore, they are more probable in some populations or ethnic groups. By comparing the presence or absence of multiple RFLPs, a likely match with a particular ethnic group, or even smaller subgroups, becomes possible.

Because RFLPs sample a limited number of sites that are somewhat distinct, it is possible that such a DNA test will identify two people who have

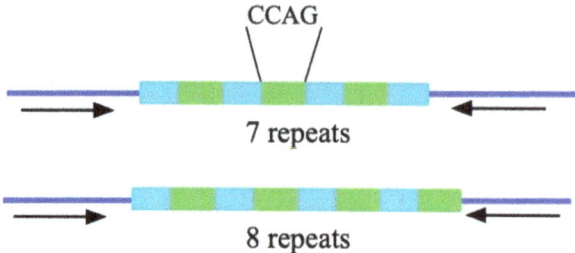

Fig. 9.18 Short tandem repeats. In the two samples, blue + green boxes contain the same short repeat sequence

almost the same patterns, so that they can discover a distant relationship, such as being distant cousins.

In addition to the procedure just described, a newer methodology is becoming more widely used. This is based on the frequent appearance of *short tandem repeats* (STRs; see Fig. 9.18). When such STRs occur occasionally in a DNA template, the DNA polymerase will frequently misalign and add an extra repeat in the DNA being synthesized. As with RFLPs, this most commonly occurs in noncoding regions, and therefore has no negative consequences.

STRs are sequences of 2–4 bases (Fig. 9.18) that are repeated up to 50 times. DNA polymerase introduces errors in copy number when copying repeat sequences, resulting in variation between individuals in the number of times a given sequence is repeated. For identification purposes, the FBI requires amplification and analysis of 13 repeat sequences, as well as a marker that determines the sex of the person providing the DNA.

In the United States, 13 core STR loci have been selected for forensic studies as standards by which an individual genetic profile can be generated. For tests involving paternity or rape (the large majority of DNA tests), a subset of Y-chromosome STRs are routinely used. Standard kits, including probes for the most common STRs, make it possible to examine 10 or more such sites in a single DNA test.

10 Mutations and DNA Repair

While DNA damage, or mutation, is generally deleterious, it may also benefit a species, as the production of new phenotypes is critical to the survival of a species, in order to be flexible in adapting to changes in its environment over long time periods. In nature, new phenotypes are produced by *chromosomal*

Table 9.4 Sources for mutations

1. Physical and chemical mutagens (= 10^5–10^6 bases/cell/day)
a. Chemical mutagens—environment
b. Radiation damage—environment
i. UV (sunlight)
ii. Ionizing, such as X-rays, etc.
2. Naturally occurring mutations
a. Point mutations—*DNA polymerase* errors
b. Insertions and deletions producing frameshifts
i. Intercalation of chemical mutagens—environment
ii. Viruses

recombination, DNA polymerase errors (see Table 9.4), and by *mutation* or change in the DNA structure induced by environmental mutagens. These are natural events, the universal source of variation critical for adapting, by natural selection, to environmental change. A *mutation* is a stable, heritable alteration in the base sequence of an organism's DNA, and may have a variety of consequences:

10.1 Autosomal Cell

1. Mutation occurs in a "silent" position (e.g. intron) = no effect.
2. Defect can be balanced by a different enzyme/protein.
3. Defect is significant, and the cell dies.
4. Defect is in a regulatory protein, and the cell becomes cancerous.
5. Defect → important malfunction in 1–2 proteins/cell/cell division → lowered function → *aging*.

10.2 Reproductive Cell (Sperm or Oocyte)

1. Mutation occurs in a "silent" position (i.e. intron) and simply produces a single nucleotide polymorphism (snp), so that descendants will have a very slightly altered DNA at a few locations.
2. Defect prevents fetal development → miscarriage.
3. Defect causes an important malfunction = genetic disease.

Examples of how the change of a single base in a codon can produce different results are shown (Fig. 9.19). Mutations at the third position do not usually change the amino acid originally specified. Changes at the first or second

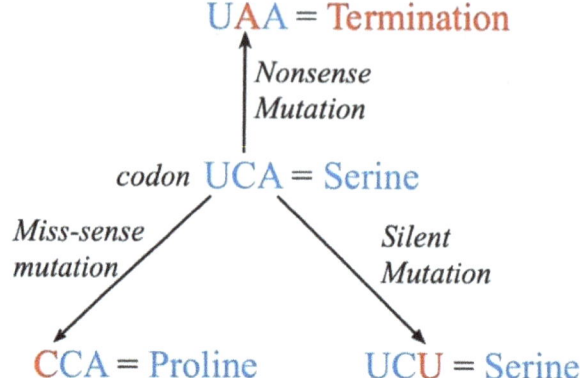

Fig. 9.19 Three possible point mutations in the same codon

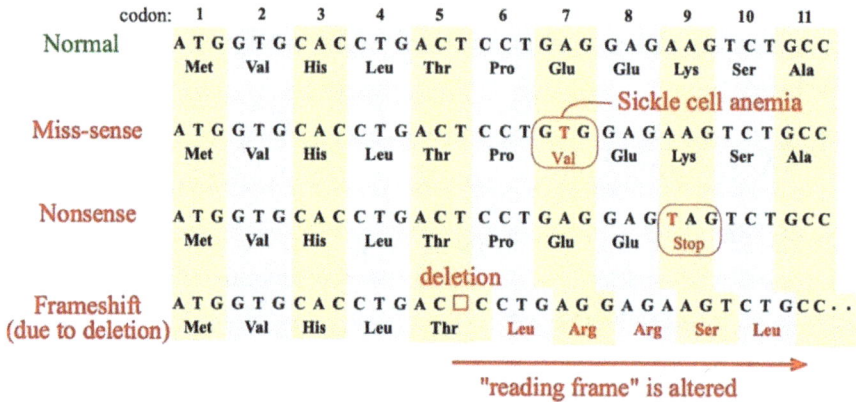

Fig. 9.20 Examples of mutations, with the DNA of β-hemoglobin. Only the first 11 codons are shown. The second row has the mutation at amino acid #6 in sickle cell anemia. The last row has a deletion (red square) in codon #5 that produces a frame shift

position always produce a change in the amino acid and can sometimes, as in this example, produce a stop codon (Fig. 9.19).

Other examples of mutations are illustrated in Fig. 9.20, using the gene for β-hemoglobin as an example. While proteins normally contain about 200 or more amino acids, for simplicity only the first 11 codons are shown. We then see how a single base change in codon #6 replaces the normal glutamate with valine. Although valine is hydrophobic, it remains on the surface of this globin protein because there are enough other hydrophobic amino acids that determine this protein's folding to make the normal folded structure still energetically favorable.

Table 9.5 Mutation rates

Species	Genome size (bp)	# cells	Polymerase fidelity (errors/bp)	# mutations/ Cell division	# mutations/ Lifetime
E coli	4.6×10^6	1	$\sim 1.0 \times 10^{-9}$	2.5×10^{-3}	2.5×10^{-3}
H. sapiens	3.2×10^9	1×10^{13}	$\sim 1.0 \times 10^{-9}$	6	$\sim 5 \times 10^{15}$

The third example shows a mutation that produces a stop codon. Even though the DNA sequence continues, when the mRNA made from this gene sequence is translated, the ribosome will stop at codon 9, producing a very small peptide with no function.

The last row shows an example of an *indel*, the abbreviation for *in*sertion/*del*etion. In this case, we have a deletion. This results when the polymerase skips one position, so that the mRNA now is missing a base (or nucleotide) in codon 5. This changes the reading frame thereafter, so that the codons now being translated will result in totally different amino acids and produce a totally different protein, which may no longer have any function.

DNA polymerases are the most accurate enzymes ever measured. While there are some variations in reported error rates, the values in Table 9.5 are fairly standard. The values for the human enzyme mean that the enzyme makes 1 mistake (adding the incorrect nucleotide) for every 1 billion nucleotides added.

It is difficult to comprehend such numbers. So let us imagine a typist. While typing, I personally make about one typo per paragraph, or 4 or 5 per page. A good typist might make only one error per 8 pages, and at about 250 words per page and 5 letters for an average word, one error per 10,000 letters. An excellent typist might be ten times better: one error per 100,000 typed letters.

Now let us imagine that this expert typist had to retype the Oxford English Dictionary (the OED to users), which is the most comprehensive dictionary of the English language. The entire dictionary contains almost 140 million words, or about 700 million letters. Our expert typist would now have to retype the entire contents of the OED, not once, but almost one and a half times, while making only one single error for this entire endeavor. Our *DNA polymerase* is truly amazing.

Actually, the above description was simplified a little to impress readers with the accuracy of DNA replication. We have several enzyme activities that work in concert. One of them has a 3′–5′ *exonuclease* activity. If an incorrect base pair forms, it will not be as stable, giving the polymerase a temporary pause, and this exonuclease domain can detect that and then use its exonuclease activity, going backward from the main direction of DNA synthesis, to remove the incorrect base.

With a genome containing 3.1 billion base pairs, or 6.2 billion nucleotides in both strands, the DNA polymerase during normal DNA replication would only make about 6 errors for every cell division. An average adult body contains 30 trillion cells. Less than half of these are routinely going through cell division. In an adult, almost no brain cells divide, and many other tissues, such as muscle, are normally quiescent.

The cells that must constantly regenerate themselves are those exposed to danger. These are the skin cells, which have exposure to UV radiation and also to possible toxic chemicals (Table 9.4). Similarly exposed are the cells of our entire digestive tract. People are not always aware of what is included with the foods that they chew and swallow. Therefore, these types of tissues are the most likely to become damaged and have the most constant need for being replaced with new cells. But even these tissues generally have only 60 cell divisions in a lifetime, and then are no longer able to divide; cells become too old.

Condensing the math in the paragraph above: humans have about 15 trillion cells that intermittently go through cell division, and the polymerase makes 6 errors per division, or about 90 trillion errors per year. Even though the DNA polymerases are amazingly accurate, they would still make about 5 quadrillion errors in a lifetime. Why are we not quickly dead because of so many mutations?

Let us review what we learned in the pages above.

10.3 Potential Mutations

98.5% occur in an intron = no harm, mostly (some intron sequences code for miRNAs, and these are important for regulation).
0.5% are at the third position of codon = silent mutation = no harm.
∴ 1% (first or second position) of all mutations cause a change in amino acid.
Further: 50% of amino acid changes are conservative = no problem.

Examples: replacing one hydrophobic amino acid with another one; or replacing one acidic (or basic) amino acid with another one.

Last: only mutations in germ line cells (sperm or ova) have permanent effects for offspring. But mutations in many tissues may lead to cancer.

Fortunately, we have additional DNA repair enzymes that routinely monitor DNA for having correct base pairs and remove and repair many of these errors. An example of DNA repair is depicted in Fig. 9.21, where an adenine base has been damaged and removed by *DNA glycosylase*. This name may appear incorrect—it suggests cleaving a sugar, but the bond broken is between the damaged base and the ribose (a sugar) to which it is attached. An *endonuclease* that detects a missing base then recognizes this spot and cuts the sugar-phosphate backbone (see Fig. 9.1).

This is followed by an *excision endonuclease* that removes about 27 nucleotides. At this point, the normal *DNA polymerase* can attach at the end of the 5′ segment (beginning of the gap) and add back the excised nucleotides, using the complementary strand as a template. A small gap remains at the end of

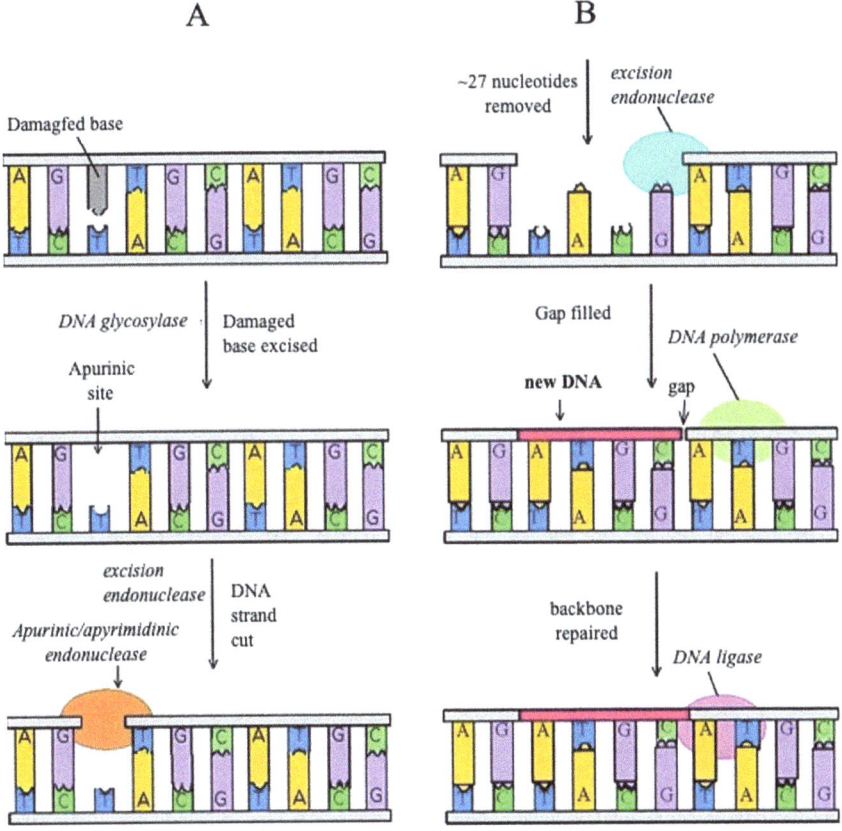

Fig. 9.21 DNA mismatch repair. (a). A damaged adenine base is detected and excised. (b). After the segment with the damaged base is removed, *DNA polymerase* fills in the missing nucleotides, and the gap is closed by *DNA ligase*

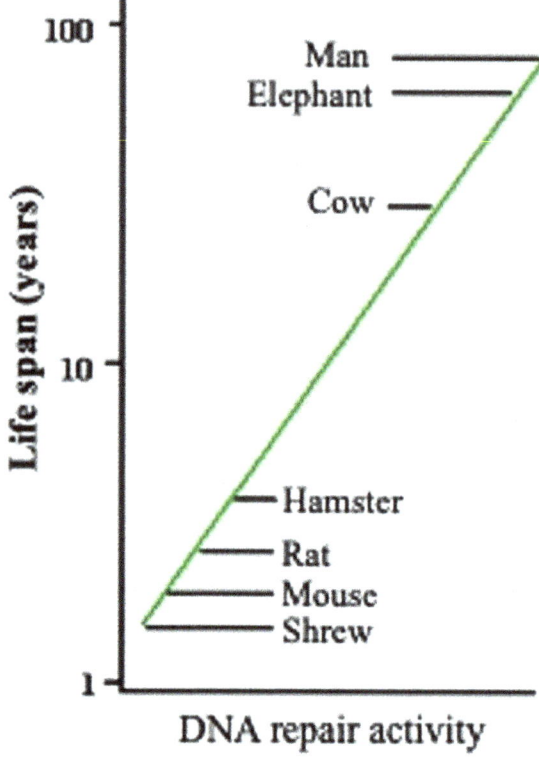

Fig. 9.22 DNA repair activity was measured in fibroblasts from the various species

this new segment of DNA, and *DNA ligase* then joins the two adjacent segments.

Again, one might be amazed at how harmoniously coordinated the six enzymes involved in this repair are. But it results simply from many of these enzymes always being in the nucleus, and when they encounter a site where they can bind, they do so and perform their normal function. The other reason why a process such as this repair can proceed so smoothly is that it is all happening in a very small space. These enzymes are constantly floating inside the nucleus, close to the DNA, and when an appropriate binding site becomes available, they can bind in less than a second or so and then automatically catalyze the reaction for which they have evolved.

Species that have better DNA repair activity also have longer average lifespans (Fig. 9.22). This appears quite logical because this ability prevents or lowers cancer and other aging disorders. A somewhat comparable relationship has been shown for the size of an organism and its maximum lifespan, as

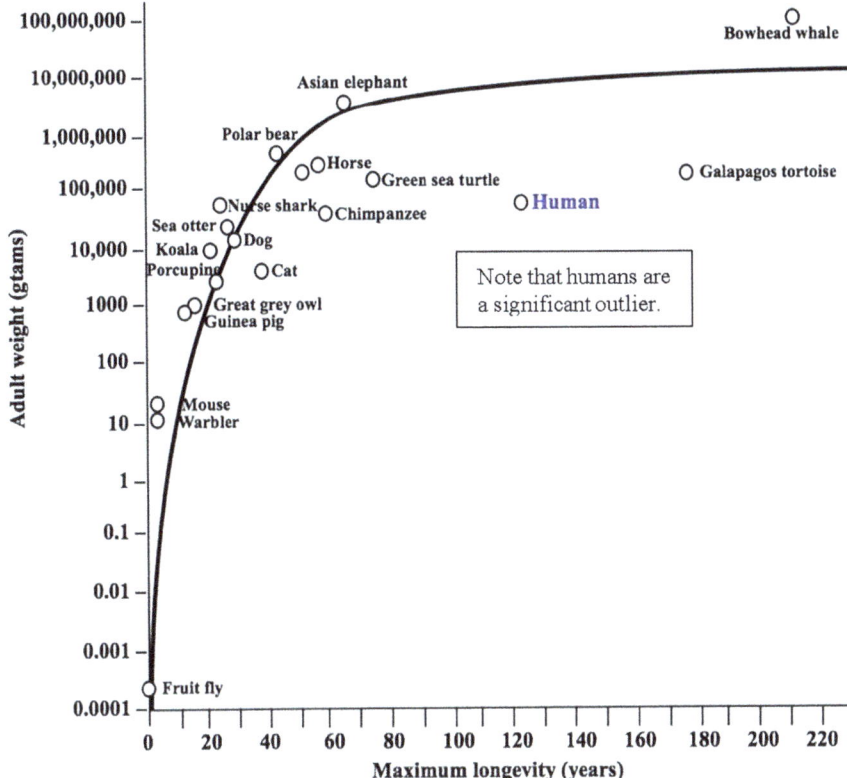

Fig. 9.23 Maximum lifespan correlates with body size for 20 different species. (Adapted from: A. Cuadra/Science (2015) 350: 1182)

shown in Fig. 9.23. Note that this graph has a log scale on the y-axis, which accounts for the curvature of the graph's line. This compression of the weight data was necessary to have both fruit flies and whales in the same graph. Because Fig. 9.22 shows increased DNA repair with body size for a few species, this may also explain the greater maximum longevity in Fig. 9.23.

If the correlation between body weight and lifespan were much better, all the small circles representing different species would be on, or very close to, the mathematical curve. Humans are an exceptional outlier. To help readers with the weight units, which are in grams to include the very small species: adults have an average body weight of about 80,000 grams or 80 kilograms = ~176 lbs. Then for our longevity, we should be far to the left on the longevity scale, where sharks and horses are, with a maximum lifespan of about 40—50 years. Even chimpanzees are modest outliers.

This longer lifespan has made it possible, and therefore has become necessary, for our slower growth rate, so that we do not mature sexually and become able to reproduce until we are in our teens. Humans need this extra time in development for our brain to mature, with its trillions of neurons. There is an emerging consensus among pediatricians, psychologists, and sociologists that humans are not mentally mature until they are in their mid-twenties, when the brain has reached full development.

Parents of a newborn must be sufficiently mature to care for an infant, while obtaining resources and making decisions about the child's care. Because we are social animals, always living in groups, much of this responsibility is handled by older members, usually women.

However, while today we have cases where girls as young as 10 or 11 have been abused and become pregnant, such early pregnancies would almost never have occurred in the stone age. Year-around nutrition would not have been adequate for girls to develop so quickly, and the ability to conceive is influenced by hormones that are sensitive to body fat. As an example, the best female gymnasts, because they are so lean, often do not menstruate until they are in their late teens.

Given the lack of hygiene in the time of the stone age, it is likely that most women were in their early 20 s before producing a surviving infant, and it would be necessary for them to live at least twice as long to help the offspring reach adulthood. It then seems that a lifespan of about 50 years, similar to chimpanzees, would have been good enough for our species to continue. The fact that we can live twice as long suggests that evolution has helped with the various enzymes needed for DNA repair and other processes, making our unusual lifespan possible for the continuation of our species in social units, where elders acquire wisdom from the experiences gained over a longer life.

11 Regulation of DNA Expression

It has already been mentioned that not all 20,000-plus genes are expressed in every cell. One of the reasons is that we have so many duplicated genes, coding for isozymes (duplicated enzymes that perform the same function). This feature has enabled the specific expression of such different isozymes in just those tissues/cells where their modified properties, in terms of K_M or k_{cat}, provide the best benefit. But also, some proteins are simply not needed in every cell. Neither the liver nor the brain has any need for the muscle protein myosin.

A major method for providing such control is by the methylation (adding a methyl group, CH_3) of DNA regions just 5′ of the coding sequence. This is accomplished by specific *methylase*s, which methylate the cytosine base in

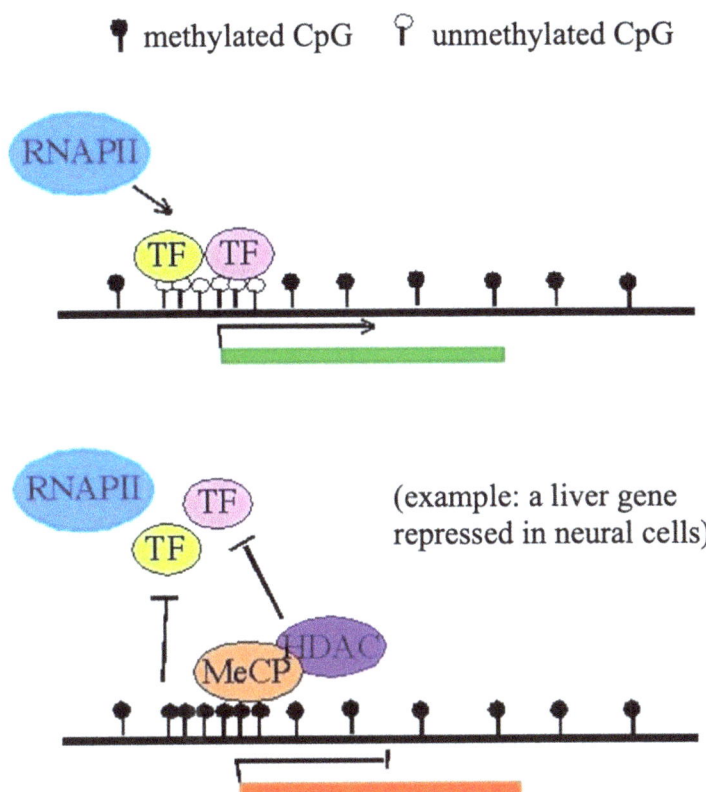

Fig. 9.24 Controlling gene expression by methylation of CpGs. RNAPII = *RNA polymerase II*; HDAC = *histone deacetylase*; MeCP = methyl-CpG binding protein. RNA transcription: ON; OFF

Cytosine-phosphate-Guanosine pairs along the same strand (Fig. 9.24). Evolution has led to the accumulation of such CpG pairs in such regions. When they are unmethylated, the adjacent sequence helps specific transcription factors (TFs) to recognize and bind at these sites. The binding of TFs then attracts *RNA polymerase II* to begin transcribing the gene into hnRNA. Methylation blocks this binding by TFs and the expression of this gene. If these regions become methylated, then regulatory proteins such as HDAC and MeCP, among others, can bind here and block the transcription factors from binding, thereby blocking RNA polymerase from binding and initiating transcription of this gene.

Resources

A history of molecular biology:
https://en.wikipedia.org/wiki/History_of_molecular_biology
Structure of DNA: https://www.ncbi.nlm.nih.gov/books/NBK26821/
DNA repair: https://www.news-medical.net/life-sciences/Mechanisms-of-DNA-Repair.aspx
Correlation between body size and longevity: https://www.sciencedirect.com/science/article/pii/S1146609X23000292
Gel electrophoresis: https://en.wikipedia.org/wiki/Gel_electrophoresis
DNA methylation: https://en.wikipedia.org/wiki/DNA_methylation

10

Uh, Oh, My Genes Are Changing: Evolution and Diseases

Abstract The process of DNA recombination is described: it is often harmful but occasionally may provide a benefit. The ancient invasion of ALU transposons in human genes has provided many possible loci for such chromosomal segments to align during gene repair and lead to swapping of DNA segments between chromosomes. Different types of recombination are described. Recombination of sets of segments within antibody genes leads to the enormous number of possible antibodies. Description of antibody structures, with variable regions as sites that bind to specific antigens. Occurrence of autoimmune diseases. The origin of human races is described, largely a result of changing productions of melanin pigments by tyrosinase isozymes, which originated in skin cells as a protection against UV radiation that could cause various skin cancers. The synthesis of vitamin D in skin cells is described and explains the limited concentration of melanin pigments that are necessary for health. A diet based on cereal grains provides ergosterol, a much less efficient precursor for vitamin D than the dehydrocholesterol found in meat, and may explain why lighter skin became necessary with such dietary patterns. An explanation of medical radiation terminology, with sickle cell anemia as an example of a helpful mutation. A process for the formation of sickled hemoglobin is described, as well as current medical therapies for this disease.

Keywords Sequence alignments • Recombination • Nucleases • Ligases • ALU segments • Transposons • Crossover • Translocation • Heavy chain • Light chain • Variable region • Double-strand DNA break • Autoimmune disease • Melanin • Eumelanin • Tyrosinase • Vitamin D • Electromagnetic radiation • X-rays • CT scan • PET scan • MRI • Ultrasound • Sickle cell anemia • Malaria • Beta-thalassemia

We have learned many things from our increasing ability to sequence DNA from ancient samples. Some of these new insights come from sequence alignments, when we have genomes of comparable sizes from different species, or just multiple sequences for a single gene. We have learned about evolutionary changes in many animals, and of course in humans. We have developed a better understanding of how many genes are actually regulated. We have learned where a specific mutation has led to an improvement in the protein coded for by a gene, or alternatively has led to it losing some function or becoming completely missing.

1   Recombination of DNA Segments by *Nucleases* and *Ligases*

Parental DNA duplexes align at sites of extensive sequence homology (Fig. 10.1), and new DNA molecules are formed when homologous segments break and rejoin. There is relatively little specificity as to the site at which the actual crossover occurs. Recombination normally occurs only between very similar base sequences. For example, recombination does not occur between sex chromosomes (X, Y) and autosomal chromosomes.

The corresponding strands of two aligned homologous DNA duplexes are nicked by an *endonuclease* (an enzyme that cuts inside DNA). The nicked single strands cross over to pair with the nearly complementary strands on the

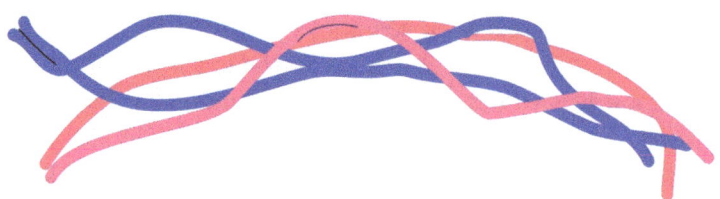

Fig. 10.1 Depiction of two chromosomes (two different colors), each with two DNA strands that can temporarily separate and then become misaligned during meiosis. Each chromosome has two chromatids (same color). Recombination may occur at crossover where strands from different chromosomes are adjacent

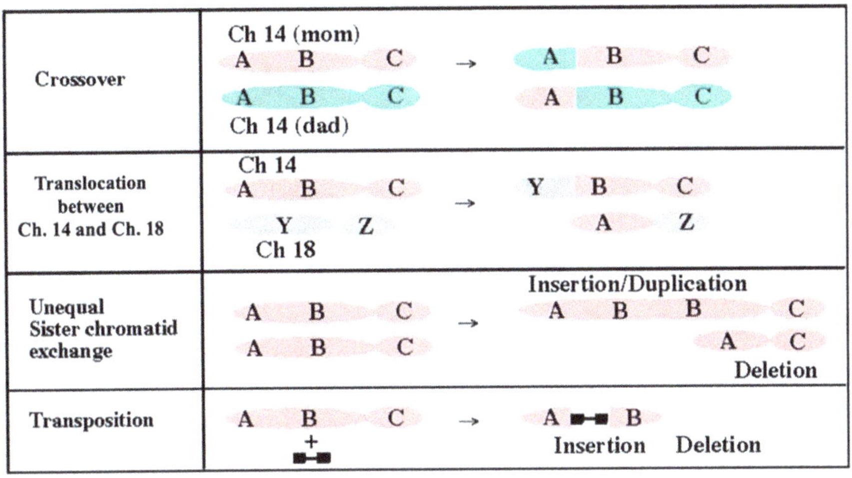

Crossover	Ch 14 (mom) A B C A B C Ch 14 (dad)	A B C A B C
Translocation between Ch. 14 and Ch. 18	Ch 14 A B C Y Z Ch 18	Y B C A Z
Unequal Sister chromatid exchange	A B C A B C	Insertion/Duplication A B B C A C Deletion
Transposition	A B C +	A B Insertion Deletion

Fig. 10.2 Examples of chromosomal recombination

homologous duplex. The nicks are then sealed to form a recombinant molecule. The resulting DNA molecule has a nucleotide sequence derived partly from one parental DNA molecule and partly from the other (Fig. 10.2).

Recombination is normally harmful, but it may occasionally be beneficial. Many enzymes are required to temporarily stabilize partially recombined DNA and to cut and rejoin the DNA. Since the construction of this enzymatic machinery consumes considerable amounts of energy and materials that the organism could use in other ways, recombination must be essential for survival. Two types of recombination are possible, depending on the nature of the original break in the DNA.

The human genome contains many small transposons, such as the sequence ALU, of which there are more than one million in our genome. This transposon is named for the restriction enzyme from the bacterium *Arthrobacter luteus* (*A. luteus*, or ALU), which cuts at this sequence. Because it occurs so frequently, DNA segments on separate chromosomes that are near each other during meiosis can misalign at ALUs, and when the enzymes clip and rejoin them after repair, they may have moved to the wrong chromosome (Fig. 10.3).

2 Antibody Diversity

We tend to think of alterations in our DNA as being generally bad and therefore undesirable. However, there is one physiological process that depends on manipulating certain specific genes routinely. This is our immune system with its ability to generate antibodies against almost any type of foreign agent.

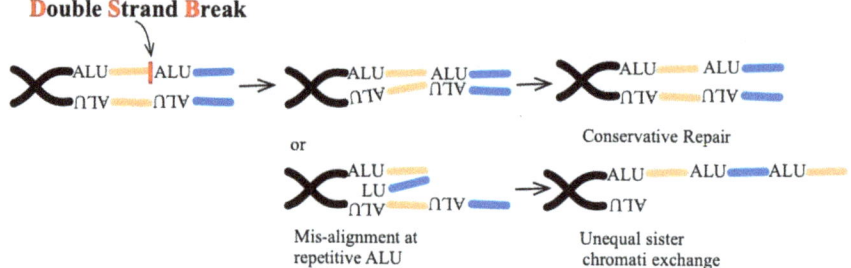

Fig. 10.3 Mechanisms for recombination with unequal exchange. Our chromosomes have many repetitive DNA sequences, such as "ALU," which occur so frequently and can result in misalignment of DNA strands, facilitating crossover of DNA segments during DNA repair

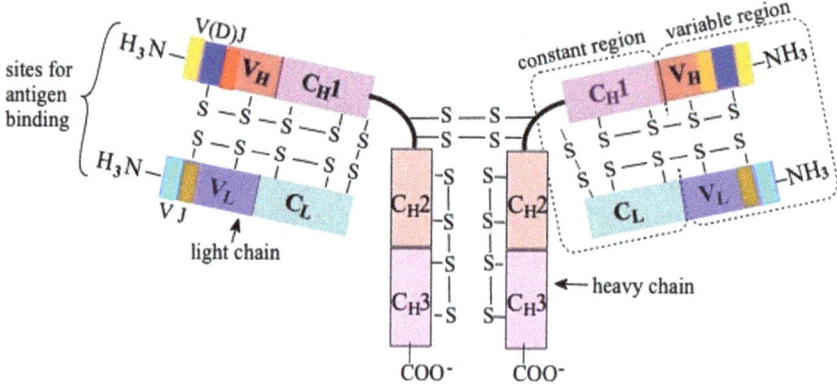

Fig. 10.4 Structure of an antibody with heavy and light chains. The top left and top right ends have only variable segments, so that many different binding sites can be produced. Segments of the two protein chains: V_H Variable Heavy chain, C_H1 Constant Heavy chain 1, C_H2 Constant Heavy chain 2, C_H3 = Constant Heavy chain 3. V_L = Variable Light chain; C_L = Constant Light chain

Let us first start with some vocabulary. An *antibody* is a protein "body" made against (anti) some infectious agent. An *antigen* (*anti*body-*gen*erating) is any molecule or part of a cell, such as a surface membrane protein, to which some variant of our antibodies can bind. We are not equipped at birth with a huge number of antibodies. We do have a subset of genes that code for the proteins that form an antibody (Fig. 10.4).

This figure shows a schematic presentation of a typical antibody molecule, which contains two heavy chains and two light chains. Each protein chain has a variable region at its N-terminus, where a variety of the different segments portrayed in Fig. 10.5 can be rearranged or recombined to produce a truly

10 Uh, Oh, My Genes Are Changing: Evolution and Diseases

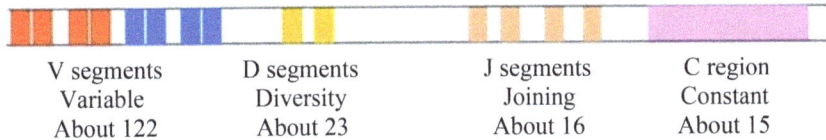

Fig. 10.5 Arrangement of immunoglobulin gene segments. These segments are actually on three different chromosomes and are combined here to show the total numbers for these different parts of an antibody

large number of slightly different binding pockets. This process is known as *V(D)J recombination*, with "D" in parentheses because only heavy chains contain a D segment (Fig. 10.4).

How big is this number? To help readers understand how such calculations are done, let us first consider a simple example of a small family with three children. How many ways could the children be in sequences such as girl, boy, girl and boy, boy, girl, etc. from oldest to youngest? Let us consider the first case: girl, boy, girl. The chances for the first child to be a girl are 50:50, or ½. The same for having a boy as the second child. Then, for any specified sequence of gender for each child, the odds are: (1/2 × 1/2 × 1/2 = 1/8). This also tells us that there are 8 possible different arrangements for the genders in this birth order.

If we now examine Fig. 10.5 and see how many different gene segments there are that can be variously recombined to make the very ends of the heavy and light chains—denoted by vertical bars in Fig. 10.4—the calculation would be:

122 V × 23 D × 16 J × 15 C = 673,440 combinations for the heavy chain.

Now light chains do not have D segments, so the calculation would be:

122 V × 16 J × 15 C = 29,280 combinations for the light chain.

Now we multiply these two results by each other to obtain almost 20 billion variations in the pocket for binding an antigen.

All 20 billion different types of antibodies will not be available in our blood at one time. They continue to be made in a random order, but if one or perhaps a few make contact with an antigen and bind to it, this will produce a feedback loop, resulting in that particular set of gene segments continuing to be made until the infecting agent, containing the antigen that is being recognized by our antibodies, is gone.

Table 10.1 Autoimmune diseases

Disease	Frequency (%)	Symptoms
Crohn's disease	Rare	Intestinal wasting
Diabetes type 1	~1	Hyperglycemia
Lupus erythematosus	~0.5	Many
Multiple sclerosis	Rare	Many
Myasthenia gravis	Rare	Muscle wasting
Rheumatoid arthritis	~0.7	Joint pain

A significant number of antibody genes are turned off during infancy because these happen to interact with proteins that are normal within an infant. Such interactions could lead to *autoimmune disease*, meaning that the immune system is becoming harmful to the body's own cells. Some examples are in Table 10.1.

Separately, and usually with advancing age, this process makes some errors, leading to translocation within or between chromosomes. Such errors contribute to the development of lymphomas and leukemias.

How is recombination accomplished? The cleavage of DNA is a two-step process and results in producing a double-strand break (see Fig. 10.3). The enzymes that perform this function are *recombinases*, known as RAG1 and RAG2 (Recombination Activating Gene). These enzymes bind and cleave the DNA at specific recombination signal sequences (RSSs) that flank each V, D, and J gene segment. Recombinases are complex, and they also have a function for reconnecting the two segments after the intervening DNA section has been excised. This then leads to a recombined novel sequence expressed in the hnRNA, and therefore in the new antibody protein.

3 Racial Differences Are Only Skin Deep

Most readers are aware of the "out-of-Africa" theory for human origins and dispersal around the globe. *Homo sapiens* (Greek for wise man) first separated from earlier, more ancestral hominids between 300,000 and 250,000 years ago. Time periods for events that far in the past are never very exact. Hominids are all primates belonging to the genus *Homo*, and among our ancestors are *H. habilis* (from Latin: able man) and *H. erectus* (erect man or upright man). It is thought that the earliest hominids developed upright posture, as many apes have, to enable them to move more easily in trees, and also have hands instead of paws to make grabbing branches and holding food easier.

10 Uh, Oh, My Genes Are Changing: Evolution and Diseases

A more erect posture, normal in modern humans, became useful in running long distances. It is thought that this was helpful in chasing herbivores that were potential sources of meat. Although these animals can all run faster than humans, they are not able to pace themselves and become exhausted more quickly. With humans running more slowly, and also more continuously, the prey would be run down and killed. Consistent with this theory of man as a hunter, it would be advantageous to have minimal body hair and to have sweat glands to cool the body as sweat evaporates. This would favor the ability to do marathon-type hunts in a hot climate. All other races diverged from these early Africans.

Living in sub-Saharan Africa, our ancestors did not need any clothing, as it was normally hot or very hot. But they were then exposed completely to the *UV* radiation in sunlight, which, over extended times of exposure, generally leads to skin cancer. Animals that have fur or body hair have skin that is light pink or almost white because the fur coat protects them from the sun's *UV* rays. It is then possible, but not established, that the very earliest humans might have had light-colored skin. But survival in the constant sunlight would quickly have led to requiring more melanin in the skin cells to help protect against *UV* light. Then, the earliest successful humans should have been dark-skinned, as Africans are today. The original Garden of Eden was in Africa.[1]

The different skin tones of different populations may result from various combinations of the different melanins (Fig. 10.6). Note the enzyme *tyrosinase*, an oxidase, which is able to use the slightly different intermediates in the melanin pathways as substrates. Having seen in Chap. 1 how specific enzymes normally are, readers might be suspicious that one enzyme could perform all these chemical reactions.

4 Tyrosinases Are Isozymes

Tyrosinases are interesting enzymes. We saw in Chap. 1, where enzyme nomenclature was described, that when an enzyme name begins with the substrate but has no term for the function or chemistry, it is normally a *hydrolase*. But tyrosinases are *oxidases*.

To help understand how one enzyme might have multiple functions, let us briefly examine Fig. 10.7. There is no need to memorize this figure or worry too much about the different structures depicted. But if we start at the top

---

[1] The "Garden of Eden" made popular by the Christian Bible comes from the lore of Hebrew culture; it would have existed in northern Iraq, where Semitic peoples originated.

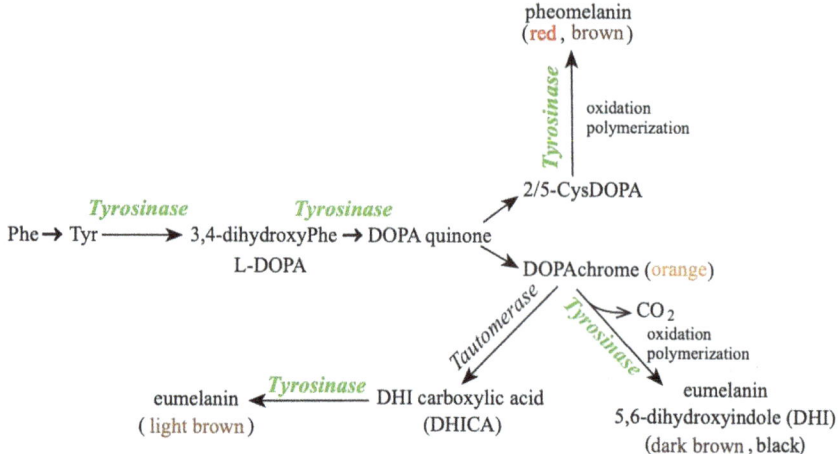

Fig. 10.6 Different skin pigments are variants of melanin

left, with the initial substrate, the amino acid tyrosine, we see an aromatic ring = the six-sided ring that forms the side chain of this amino acid. Such aromatic rings are very capable of absorbing UV light and normally absorb a limited range of wavelengths effectively. This is what gives color to any cells that have these compounds, of which melanins are more developed end products. The sequence of oxidative steps is largely centered on the two carbons that form the left side of this ring, which I have highlighted with a red dot.

The need to have such pigments that can absorb UV light goes back to the first forms of life, as this offered protection before the ozone layer had formed (Chap. 4). There was almost no oxygen available at the beginning for this protective atmospheric layer to exist. It has been found that all forms of life have the ability to make most of the various melanin pigments, and this is consistent with the observation that the tyrosinase enzymes in these different organisms have very little resemblance to each other. This suggests that they arose independently because their function was so necessary.

Figure 10.7 also indicates the most obvious color associated with different intermediates in the formation of eumelanin and pheomelanin. While the figure may appear to portray that melanocyte cells should always just make the final compounds at the bottom of this figure, it is possible for the melanocytes to contain an assortment of both intermediates plus one of the end products. This would then provide a blending of colors and explains why so many color variations are possible in human skin, hair, or eyes.

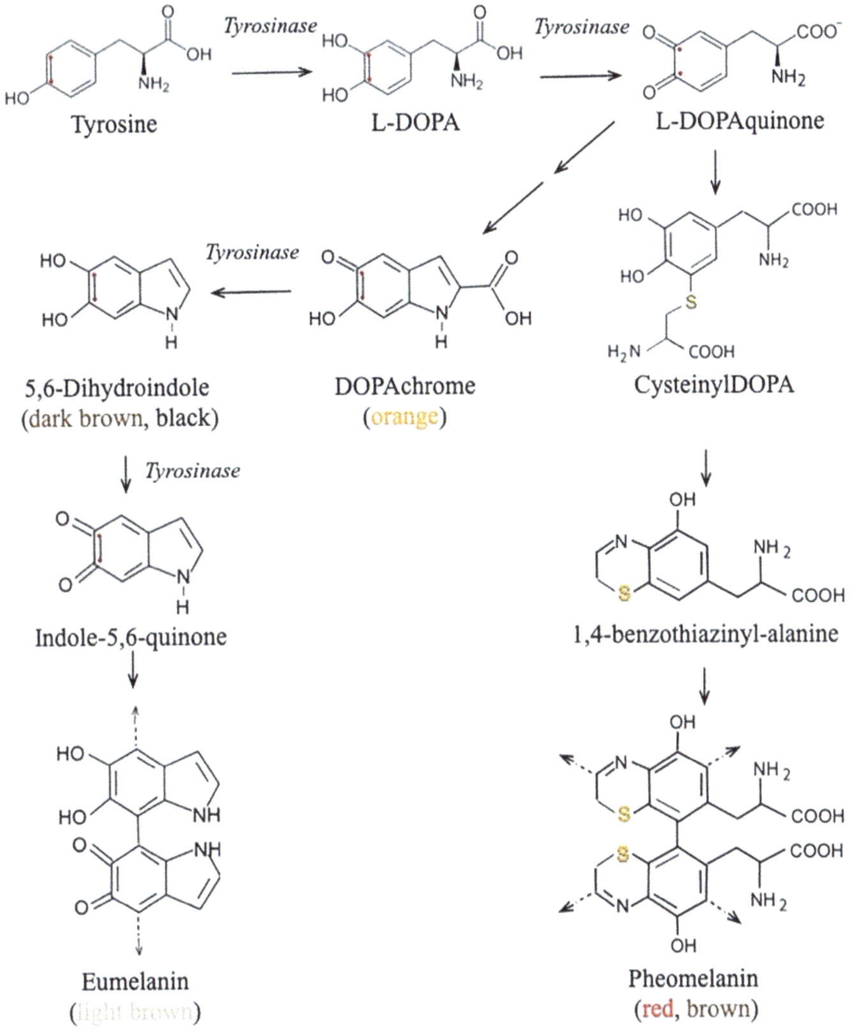

Fig. 10.7 Pathway for the synthesis of melanin pigments. Red dots indicate the carbons to be oxidized. Dashed arrows on eumelanin and pheomelanin indicate positions where the same structure will bind repeatedly to form polymers

In humans, there are four different *tyrosinases* (isozymes) that vary somewhat as to which intermediate substrate they bind best. This ability to control the concentration of both the intermediates and the end-product melanin compounds explains how the variations in color can then be regulated or controlled by regulatory proteins that influence the translation of the tyrosinase RNA. Figure 9.11 depicts a normal gene, showing both exons and introns. In this figure, we see how the process of splicing the initial hnRNA leads to the removal of the introns and the joining of the exons to form the final mRNA.

The process for transcribing the tyrosinase gene is somewhat different because it uses *alternate splicing*. It appears that humans have a single gene for tyrosinase, but that this gene contains at least four separate/different exon regions coding for similar but different regulatory domains. Depending on the influence of the regulatory protein in the melanocytes, only one of the four regulatory domains is included in the original hnRNA, and therefore in the final protein being made. It is then possible to have four slightly different tyrosinases, each with a somewhat different affinity for the potential substrates depicted in Fig. 10.7.

Most likely, there was just one form of the tyrosinase being made in the earliest cells, but the benefit of having variants of this enzyme must have favored the duplication of the regulatory region within the gene, plus the ability to use alternate splicing so as to produce four distinct isozymes. Readers should appreciate that the procedure just described is somewhat unusual. The normal process for producing isozymes is for the gene to be duplicated several times and then to have the duplicated versions undergo modest mutations, thereby providing some benefit with slight differences in their affinity or rate with the different melanin pathway intermediates.

While the need for melanins in our skin as a protection against cancer is fairly logical, we should also appreciate how evolution can lead to new functions for established compounds. Not only is melanin synthesis important in our skin cells, but it also became functional in hair cells and cells of the iris, as these features produced survival benefits.

Most readers are familiar with the concept of evolution, first expressed by Charles Darwin, as the survival of the fittest. That naturally suggests that evolution should act on genes that improve strength, speed, or intelligence. But evolution also selects for genes that make a person more sexually attractive. After all, people who have such features are more likely to attract a mate and then have children. Whatever genes are responsible for such enhanced good looks will continue in the children. This may help to explain how blond, red, or auburn hair became more widespread in Caucasian peoples. And it may have also favored the appearance of blue and gray eye colors.

5   Vitamin D Synthesis Occurs in Skin Cells

Why did Africans leave their homeland? Obviously, humans, like almost all animals, respond to changes in, or lack of, food and other resources, and roam farther afield in search of better environments. There is archaeological evidence that this happened with *H. neanderthaliensis* and *H. denisovan*, as these species left Africa much earlier than the later *H. sapiens*.

However, about 74,000 years ago, there occurred one of the greatest volcanic explosions with the eruption of Mt. Toba in Indonesia. From the size of sedimentary layers in rock formations, it has been estimated that the volume of ash and debris would have lasted in the atmosphere for perhaps two years, obscuring sunlight enough that most plants would have died, and then the herbivores that ate the plants, leading to near-famine conditions for humans.

Using genealogical DNA sequence alignments, it has been possible to trace the ancestry of all human races back to Africa, about 65,000 years ago. The dates for the Mt. Toba explosion and for the exodus from Africa both have some uncertainty, but are close enough, and in the right temporal sequence, to suggest that this was the cause of this great emigration event, because the loss of so many plants might take thousands of years to recover, and then so did the reappearance of herbivores as important food sources. Archeological and anthropological data show that many Africans by this time survived in coastal regions, where shellfish, kelp, and fish could be obtained easily. It would then be easy for Africans along the east coast to simply wander up to the Sinai Peninsula and continue northward into Europe. Also, at this time, the northern hemisphere was in an extended ice age, meaning that large parts of Europe and North America were covered by extensive ice sheets or glaciers. Therefore, sea levels would have been lower, and the Red Sea much lower and narrower, so that it became possible to cross it from Ethiopia, reach Yemen, and then continue into Asia.

It is well known that most humans outside of Africa, except Central India and Southeast Asia, are not dark-skinned. Why did they change? That Africans, after leaving their home continent, led to descendants with lighter skins was so well established that when the journal Science published an article with the first genomic DNA sequence obtained from Neanderthal bones, they added an ironic photo to emphasize this accomplishment (Fig. 10.8a). In this photo, a normal young American man is looking at his ancient cousin to the right. The first Neanderthal relics were discovered in Germany, and that is how the species was named Neanderthals (from German: Neander Thal = Neander valley). So, clearly, these ancient cousins had to be Caucasian.

A Science (2010) 328, 680　　　　**B** Science (2024) 385, 132
(source: License Number 5861980086067)　(source: Order license ID link:1522699-1)

Fig. 10.8 Changes in how a journal presents images of Neanderthals

Also amusing is that the old Neanderthal, living in the stone age without tools having sharp edges, somehow shaved the forepart of his scalp and much of his face.

Fourteen years later, in another article on the genetics of Neanderthals, the editors presented a computer-generated image (Fig. 10.8b) showing believable Neanderthals. Why was the image at left wrong? The answer involves the final steps of vitamin D synthesis, requiring activation by UV light, that are partly finished in skin cells just beneath the upper skin layer.

There are two precursors for vitamin D in foods that can be activated by *UV* rays, and then further processed by *hydroxylase enzymes* to become the functional $1,25\text{-}(OH)_2$-vitamin D (Fig. 10.9). The numbers refer to the carbon atoms in the structure where hydroxyl groups must be added by these enzymes. Vitamin D_2, ergocalciferol, is a less effective precursor, and we require ten times more to make the final active vitamin.

This becomes more difficult in northern climates where days may be shorter, and people wear more clothing. This was not a problem as long as people, as in Africa, had been hunter/gatherers, and so ate adequate amounts of meat or fish to provide the better precursor, cholecalciferol (Vit. D_3).

This worked fine, even for Neanderthals, but approximately 10,000 years ago the last ice age began to end, and Neanderthals had become extinct by then. By 7000 years ago, with the increasing human population, we had the beginning of widespread agrarian cultures growing all the standard cereals: maize, corn, wheat, rice, etc. And as urban cultures expanded, the daily diet of poorer people became heavily dependent on such grain sources. These grains made cities and armies possible because the grains could be stored and

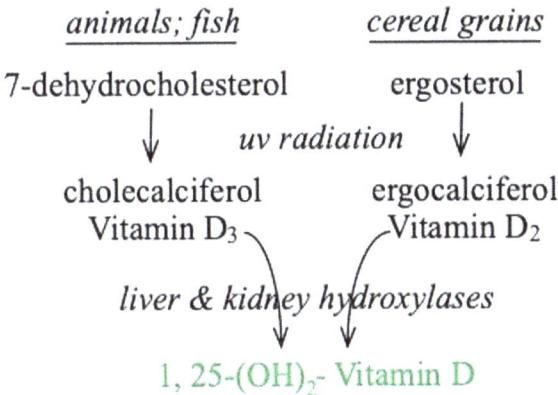

Fig. 10.9 Activation of vitamin D precursors

therefore available throughout the year, and some men were no longer required to hunt or gather food. They could become laborers or warriors.

For darker-skinned people, this resulted in vitamin D deficiencies. Natural selection would then have favored the survival of people having lighter skin. It is probable that within a few thousand years, lighter-skinned people would have become the norm. The active form of the vitamin is actually a steroid hormone and is necessary as a transcription control factor for the activation of several hundred genes. Therefore, any significant deficiency in vitamin D has serious clinical consequences.

6 Medical Radiation Jargon Explained

We are all somewhat familiar with the various scanning procedures used by hospitals and clinics to obtain better information about a patient's condition. While we all have a sense that radiation can be harmful, we still submit to having these procedures done to us because they can be so helpful to the doctor or dentist in deciding on some treatment.

Radiation at the short-wavelength (= higher frequency) end of the electromagnetic spectrum—high-frequency ultraviolet, x-rays, and gamma rays—is ionizing. Ionizing radiation has sufficient energy to remove electrons that are tightly bound from their orbit around an atom. This causes that atom to become ionized and can alter its chemistry.

Frequently, this leads to single- or double-strand breaks in DNA. Lower-energy radiation, such as visible light, infrared, microwaves, and radio waves, is not ionizing. Table 10.2 shows the various electromagnetic radiations in our

Table 10.2 Electromagnetic radiation

↑

Gamma rays
X rays
Ultraviolet light
Visible light
Infrared light
Radio waves
Heat wave

---

environment, with the most energetic, dangerous, at the top. While gamma rays are the most dangerous, they are also the most uncommon. They are normally caused by suns going nova, but they can also penetrate rock.

Radiation received from medical/dental imaging accounts for at least 50% of all radiation exposure in the United States. Despite standards from US Radiologists, surveys suggest that many hospitals/doctors almost never follow "established guidelines."

Types of medical radiation:

X-rays: Energetic radiation, usually focused on a small area.

CT scan (computed tomography; tomography from Greek *tomos* for part, i.e. section): X-rays, but over a larger area and more time. By repeatedly focusing on thin, but deeper layers, a computer can reconstruct a 3D image of the scanned area. CT scans alone account for at least 25% of all radiation exposure in the United States.

PET scan (positron emission tomography): A gamma-ray-emitting compound is swallowed and goes into the body. Gamma rays are more energetic than X-rays. Gamma rays and positrons are produced together by the swallowed molecules, and gamma rays are detected.

MRI (magnetic resonance imaging): A machine generates a strong magnetic field that momentarily affects the spin of electrons in molecules of tissues; this can be detected within the magnetic field. No ionizing radiation.

Ultrasound: Uses high-frequency broadband sound waves in the megahertz range. No ionizing radiation.

7   Sickle Cell Anemia: A Helpful Deleterious Mutation?

We need to balance our understanding of mutations in that they most commonly are neutral or bad, but occasionally they can provide a benefit. After all, all evolutionary advances resulted from some occasional mutations that provided a change in how some enzyme or protein functioned, and thereby altered some aspect of anatomy or of physiological function for the better.

An example of a mutation that can be both deleterious and helpful is provided by the gene for beta hemoglobin. Two types of alterations have occurred and are largely interpreted as favoring survival in areas where the disease malaria is prevalent (Fig. 10.10). *Thalassemia* (from *Thalassos* = Greek name for the Mediterranean sea) is the absence of the HbB protein, resulting from little or no expression of this globin gene. Its original appearance is unclear, but the spread of this gene has been ascribed to the armies of Alexander the Great, as his soldiers marched into Asia. The occurrence of this gene in the twentieth century is fairly comparable to the regions conquered by this army in the fourth century BCE.

The other disease is *sickle cell anemia*, which is caused by a mutation in HbB and results when an individual makes a mutant form of the β hemoglobin (β^S or HbS), when amino acid #6 is changed from glutamic acid to valine (see Fig. 9.20). After hemoglobin becomes deoxygenated, there is a lag time before polymerization occurs. During this time, nucleation of HbS fibrils

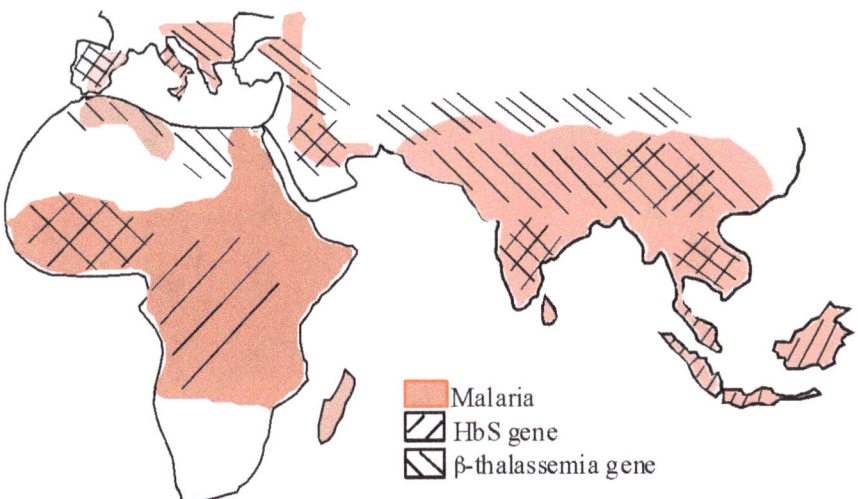

Fig. 10.10 Occurrence of red blood cell diseases and malaria

occurs until a critical size is reached. After this, fibrils form rapidly and polymerization becomes very extensive, as illustrated (Figs. 10.11 and 10.12). This process has been studied with purified hemoglobin. After the hemoglobin has been deoxygenated, the appearance of fibrils due to polymerization takes at least 50 msec (milliseconds), with an average time of about 160 msec, while the average time for a red blood cell to slide through a capillary is 300–700 msec.

How was the sickling in Figs. 10.11 and 10.12 measured? To physically measure polymerization of HbS, the device shown (Fig. 10.13) was used. A clear plastic block has a narrow passage. Two syringes connect to a mixing chamber, which leads to the channel in the plastic block. One syringe has the starting solution of RBCs (red blood cells), and the second has an appropriate buffer, which defines the final concentration of CO.

The concentration of CO that is equivalent in binding to HbA with O_2 had been determined previously. And because CO binds so much more tightly, it enabled the rapid mixing needed for these studies. A laser shines through the plastic block with the flowing solution. The laser projects an image on the screen so that sickled cells can be observed (Table 10.3).

The delay time for polymerization is very variable, due to the randomness of nucleation. For single cells, this varies from 1 msec to 100 s. 10 to 100 msec are the most common measured times, and the average equals 160 msec (for $pO_2 = 0$). At this pO_2, hemoglobin must be in the T conformation.

To measure unbound (anoxic) Hb, the cells are saturated with CO (carbon monoxide),[2] for 100% binding, and also to displace any O_2 that was in the

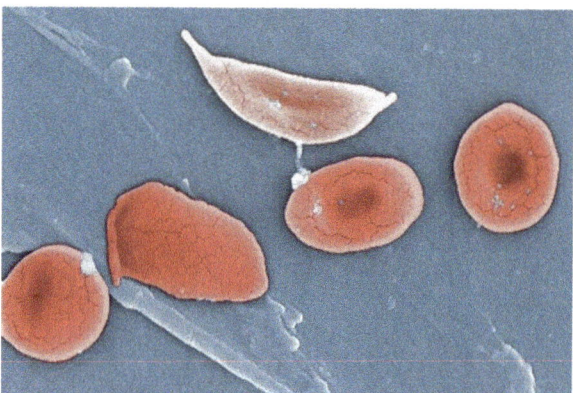

Fig. 10.11 Example of a sickled red blood cell. (Source: https://commons.wiki media. org/wiki/File:1911SickleCells.jpg#file)

---

[2] Carbon monoxide binds to the same location on hemoglobin as oxygen, and binds more tightly, making it a very toxic gas by displacing oxygen.

10 Uh, Oh, My Genes Are Changing: Evolution and Diseases

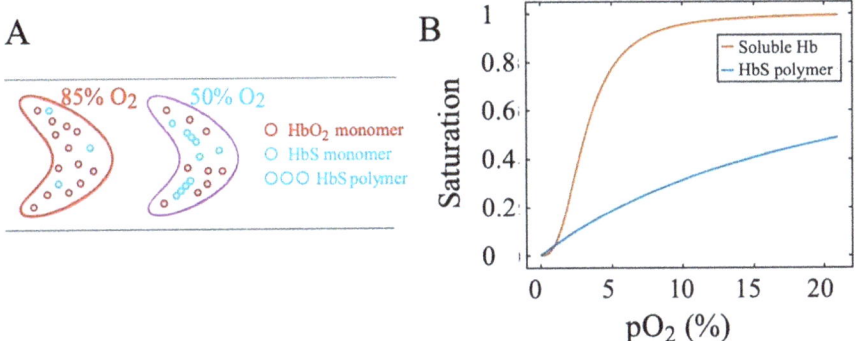

Fig. 10.12 Normal and sickled red blood cells as a function of pO_2. (Source: G. Di Caprio et al., PNAS 116 (50) 25236–25242)

cells. The solution is then pulsed with the laser for instantaneous removal of CO → zero binding, and as the solution flows past at a fixed rate, the screen shows how many cells have formed polymers. Cells whose Hb has been made empty with the laser can then be mixed again with buffer containing predetermined CO concentrations to measure the rate at which polymerization disappears.

To understand how such an apparently simple mutation can lead to the formation of fibrils, let us examine Fig. 10.14. Note how the normal, but unimportant, appearance of a small hydrophobic pocket on the surface of a β subunit is exactly positioned to accept the hydrophobic valine that is now on the surface of the mutant. The normal HbA hemoglobin has a negatively charged glutamate at this position, which would never bind into a hydrophobic pocket.

The possibility of individual hemoglobin tetramers being positioned as shown in Fig. 10.14 is facilitated by the fact that each red blood cell contains about 280 million hemoglobin molecules, or almost 95% of all the protein molecules in these cells. Being so abundant, at any time, some subset of these hemoglobins will, by chance, align as shown. In Fig. 10.14b, note how when another tetramer aligns oppositely, as shown, it will now overlap two tetramers, and as this process is repeated (Fig. 10.14c), the polymer becomes sufficiently extended. Polymers will then bundle with one another and form much larger fibrils, which are now sufficiently stiff that as the cell becomes deformed by sickling, the fibers may puncture the cell membrane (Fig. 10.15).

These data are shown in numerical detail in Table 10.3, where we see how much the solubility of hemoglobin has decreased. Observe also that

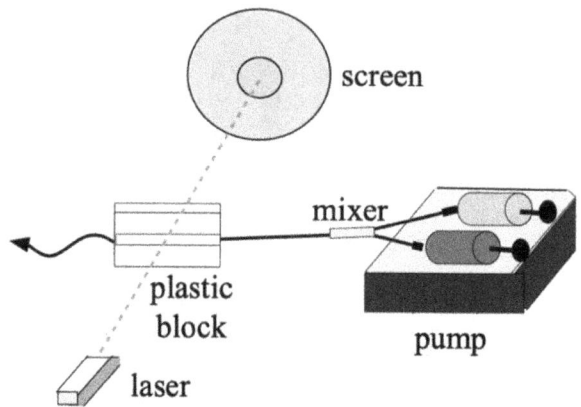

Fig. 10.13 Measuring polymerization of hemoglobin. The laser beam can remove varying amounts of bound CO, to approximate any concentration of O_2, including "0." Then a new solution of hemoglobin plus a fixed [CO] is pumped through the block

Table 10.3 Time for red blood cells to make a round trip

Part of circuit	Average time
Alveoli of lungs	0.7–0.3 sec oxygenated
Arteries	3 sec oxygenated
Capillaries of tissues	0.7–0.3 sec deoxygenated
Veins	10–20 sec deoxygenated

homozygous sickle cell people have only half of the normal concentration of hemoglobin because their cells die much faster.

The life-shortening anemia happens to people who are homozygous, having two copies of the mutant gene. If heterozygous (one normal and one mutant gene) couples mate (Fig. 10.16), then ¼ of their children will have both normal genes and therefore be likely to be infected with the malaria parasite, and ¼ will have both sickle cell genes and therefore be frequently ill and die at a younger age. But half of the children will be heterozygous and then lead a normal life.

As shown in Table 10.5, the mutation leading to sickle cell anemia was itself a fortuitous development, as sickling disrupts the cell membrane enough for the concentration of K^+ (potassium) to become much lower, leading to the parasite being unable to survive. But these few holes in the membrane can be mended, and the cells will not die. Therefore, heterozygous people will not have anemia (Table 10.5).

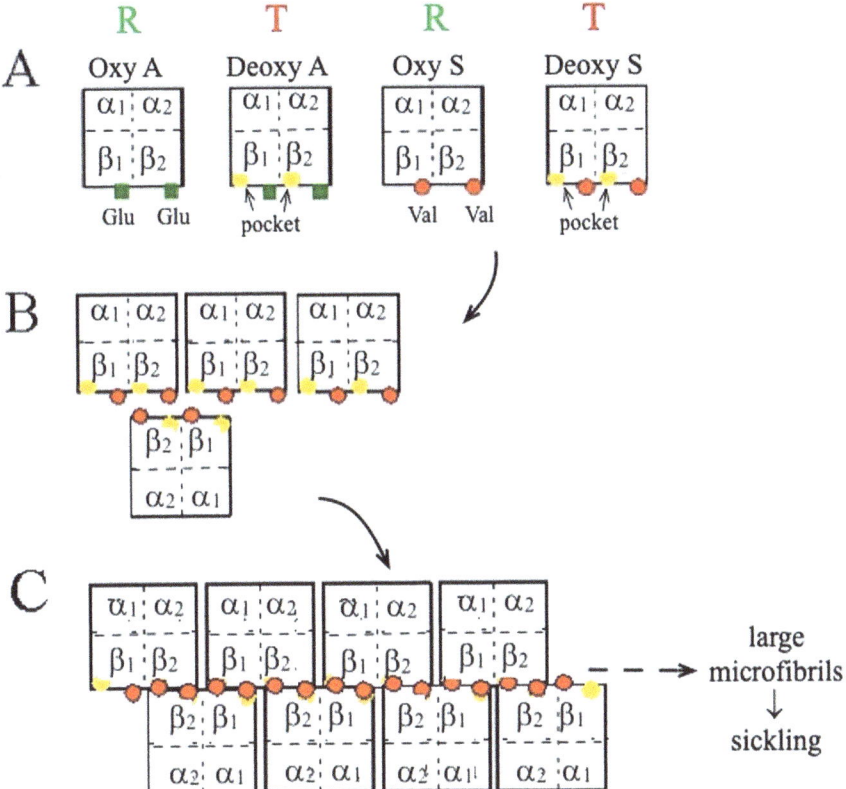

Fig. 10.14 Mechanism for polymerization of deoxy-hemoglobin S. (**a**) At left: In normal hemoglobin, the β chain develops a small hydrophobic pocket as the hemoglobin protein is deoxygenated (yellow); at right: the mutant β subunit does this also. This has no consequences. (**b** and **c**) However, in HbS the mutation, a valine, produces a hydrophobic surface residue (red) that is positioned exactly to fit into this pocket on an adjacent tetramer when these proteins are in the deoxy state

8 Treatments for Sickle Cell Anemia

However, people who are homozygous have such consistent and severe anemia that they are intermittently in pain, always weak, and their lifespan is often shortened from all the health complications. Therefore, there has been a continuing effort to find drugs that might help.

1. **Hydroxyurea** (originally just urea): Most widely used; *least toxic*, though side effects may occur. Main effect is to increase expression of the HbF gene.

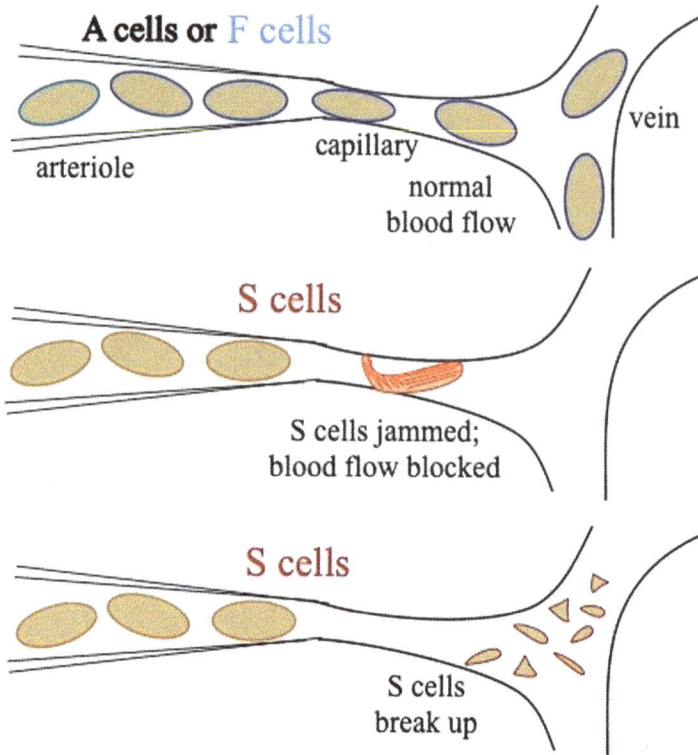

Fig. 10.15 Comparison of F cells with S cells. HbS cells are more likely to block a capillary or, because of continuing blood pressure, to break up. F cells are not as efficient as normal HbA at transporting oxygen for adult needs

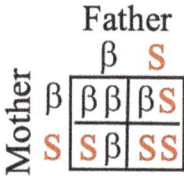

Fig. 10.16 Gene pattern for children

2. Cyanate: Carbamylates proteins; lowers the p50 so that there is less sickling. It was not publicly accepted, perhaps because the name sounded like cyanide.
3. Nitrous oxide (NO): Binding by Hb lowers the p50 by 5 torr, so less sickling. It was recently found that inhaling a nasal spray is helpful. Not yet clinically approved.

Table 10.4 _____ Hb solubility_____

Hemoglobin gene for β	[Hb]cell	$pO_2 = 0$ torr	$pO_2 = 50$ torr	pO_2 arterial	pO_2 venous
Hbβ Hbβ	60–70 g/dl	70 g/dl	70 g/dl	92 torr	39 torr
HbS HbS	32 g/dl	17 g/dl	23 g/dl	85 torr	46 torr

Table 10.5 Benefit of the sickle cell trait

1) parasite inside RBC → ↓ pH (due to increased glycolysis
2) ↓ pH → ↑ sickling → ↑ membrane permeability (holes in membrane)
3) ↑ membrane permeability → ↓ K^+ (ions leak out)
4) ↓ K^+ is lethal to Plasmodia (malaria parasite)

4. Hemoglobin gene therapy:
    (i) Only a few successful examples.
    (ii) Identify new compounds to activate the transcription factor for the HbF gene. Sickle cell anemics with 15% HbF have few symptoms.

Increased concentrations of HbF cells have proven to be helpful. The gene is almost completely turned off in normal adults, but it has been observed that sickle cell patients actually have somewhat higher levels of this protein. This higher amount may also reflect the fact that cells containing HbS tend to be destroyed more quickly.

Resources

Genetic recombination: https://www.nature.com/scitable/topicpage/genetic-recombination-514/
B cells and antibodies: https://www.ncbi.nlm.nih.gov/books/NBK26884/
Vitamin D synthesis: https://www.sciencedirect.com/science/article/pii/S209012321400023X
Neanderthals: https://evolution-outreach.biomedcentral.com/articles/10.1007/s12052-010-0250-0
Sickle-cell disease: https://simple.wikipedia.org/wiki/Sickle_cell_disease

11

Control Your Sugar: Special Enzymes Control Glucose Metabolism

Abstract Glucose evolved to be the principal sugar in mammals because it is the least chemically reactive and, therefore, the least harmful in modifying other molecules when it is at a higher concentration than other sugars. In blood, it has a concentration near 5 mM, with a range of 3.9–7 mM considered to be normal. Glycosylation of DNA can lead to mutations. Explanation of glycogen synthesis and the conversion of glycogen back to glucose, when stimulated by hormones. Glucose is stored as glycogen to minimize the water needed to solubilize many individual glucose molecules. Activation of glycogen breakdown to glucose is initiated by adrenaline or glucagon. Within liver or muscle cells, a cascade of protein kinase enzymes leads to the activation of glycogen phosphorylase, which produces the needed glucose-6-P; in liver cells, this is dephosphorylated and transported into the blood as free glucose, to be available to all tissues, but mainly to be used by the brain and red blood cells. The two key regulatory enzymes in this process are regulated by ligands that have a short lifetime, thereby controlling the duration of the process. Description of gluconeogenesis from amino acids and ketones. Description of different glucose transporters and of the four hexokinase isozymes. Lactate dehydrogenase converts NADH back to NAD^+ during amplified glycolysis. The main regulatory enzymes controlling glycolysis are phosphofructokinase, pyruvate kinase, and lactate dehydrogenase. Brief description of the citric acid cycle and oxidative phosphorylation. Discussion of energy sources during short-term starvation and long-term starvation. A brief description of fatty acids and triglycerides.

Keywords Glucose • Glycemic value • Hyperglycemic • Hypoglycemic • Glycosylation • Glycogen • Glucagon • Adrenaline • Epinephrine • Phosphatase • Protein kinase • Adenyl cyclase • cAMP • Disaccharide • Glycolysis • Hexokinase • Erythrocytes • Red blood cells • Amplified glycolysis • Citric acid cycle • Phosphofructokinase • Fatty acid • Triglyceride • Oxidative phosphorylation

Glucose is the most important sugar in our bodies. All sugars can bind to various other metabolites and chemically alter them, and glucose is the least reactive at doing this. This may explain how glucose came to be selected by evolution for its various roles, as these different functions have made it important for this sugar to be at higher concentrations than any other sugar.

Most readers are likely aware that for energy, humans can also use fats. Fats are metabolized differently from sugars. On a weight basis, fats contain more than twice as much energy (more calories) as sugars (carbohydrates or "carbs"). But sugars can be metabolized more quickly for immediate energy, and glucose especially has become the preferred energy source for our brains. That is why evolution has provided us with taste buds for sweet, and a neural response that tells us how enjoyable this taste is, so that we will seek more of these foods. Our brain knows how to look out for itself.

As will be explained soon, our bodies can also make glucose from other precursors, most commonly amino acids. Therefore, we should expect that since glucose is so important, and since its concentrations ideally need to be kept within a narrow range, we should have multiple mechanisms for achieving this.

1 Overview: The Importance of Glucose

In Chap. 3, we learned that for adults, the concentration of glucose in the blood is normally about 5 mM. This would be a person's *glycemic value*, and this number represents an average that can vary somewhat—but not too much. If the concentration of glucose in the blood is above 7 mM, that would be *hyperglycemia*, and the patient would be considered to be prediabetic.[1] If it is below 3.9 mM, it would be *hypoglycemia*. The latter is bad because then that

---

[1] In American medical clinics, blood glucose concentrations of 3.9–9.9 mM are considered normal. This may reflect that many people are overweight and prediabetic.

person would not have enough strength or endurance for normal physical activities. Symptoms include fatigue, clumsiness, faster heart rate, and seizures.

Hyperglycemia is also bad. Though it signifies abundant energy resources, the continued high concentration of glucose means that glucose will slowly react nonenzymatically with any available proteins or other molecules, and lead to organ damage involving the kidneys, the retinas, and vasculature in the lower legs/feet. If uncontrolled, this may lead to a need for surgical amputation of a foot or lower leg. The immune system also becomes less effective at preventing/fighting infections.

Let us examine Fig. 11.1. This is a schematic representation of a liver cell. Among the liver's many functions, it is a central organ for synthesizing and storing glycogen, depicted at center right. Starch, made by plants, and glycogen, made by animals, are similar polymers of glucose and serve as a storage form for this important sugar. These two types of glucose polymers both link glucose molecules together via an alpha linkage between carbon 1 of one glucose and carbon 4 of the next glucose, defined as an α(1–4) bond or linkage. Glycogen is slightly more complex, as it also has α(1–6) bonds. Therefore, these polymers are equivalent in energy, and the same enzymes can release glucose from either polymer.

An interesting, trivial fact: along with lignin, cellulose is the major structural molecule of plant fibers, especially wood. Cellulose is a glucose polymer made with β(1–4) bonds or linkages. So, if you are looking at a redwood tree, you are seeing a very large amount of sugar in a special structural arrangement. Mammals do not have any enzyme that can break down cellulose because the bond between the glucose molecules is sufficiently different. Termites, however, are quite good at doing this.

Why do we need to make things more complicated by forming polymers of glucose? This is due to the need to minimize the amount of water needed that will interact with any compound in solution. Individual glucose molecules, as in our blood, have many water molecules interacting with them. That is what it means to be "soluble." But, within a cell, there are so many molecules competing for the available water that it has become better to make large polymers that can pack together, like an iceberg in the sea, and thereby make it possible to maintain larger quantities of these important metabolites. We make large polymers of fatty acids for the same reason.

For simplicity, the glycogen complex in Fig. 11.1 has only 6 glucose molecules, but an actual polymer in a liver cell would have hundreds or thousands. The release of individual glucose molecules is accomplished by *glycogen phosphorylase*. Before this enzyme is covalently modified by *phosphorylase kinase*, it is named *glycogen phosphorylase-b*. This latter form of the enzyme is

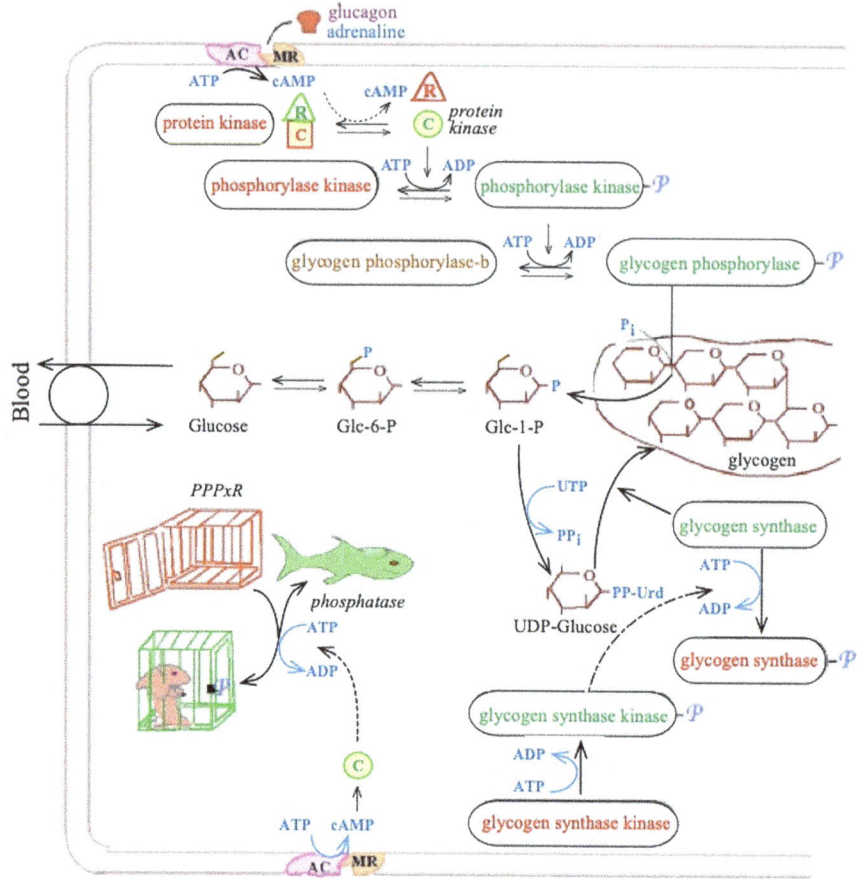

Fig. 11.1 The importance of enzyme phosphorylation in the control of glycogen metabolism: (**a**) in the membrane, AC = *adenyl cyclase*, MR = membrane receptor; (**b**) in the cytoplasm, C = catalytic subunit, R = regulatory subunit of *protein kinase*. *PPPxR* is a regulatory subunit that binds the *phosphatase*. Note that when the *PPPxR* is phosphorylated, it binds and inactivates the *phosphatase*. Allosteric changes are produced by (1) binding of allosteric ligands, or (2) covalent modification = *P* enzymes. (Illustration by Jason Traut)

not totally inactive, as we see with the other enzymes in this figure. It is moderately active because some release of glucose molecules is required almost constantly, except for an hour or so after a meal. This moderate activity guarantees that enough glucose will always be in the blood to keep our brain and RBCs functioning at a high level.

While Fig. 11.1 depicts a liver cell, most of this figure could also represent a muscle cell. Muscle is the other tissue that has the enzymes to synthesize glycogen, and then to release glucose molecules when more strenuous activity is needed. When we are at rest, our muscle cells depend more on fats for energy, saving their stored glucose for the more important high-demand times that occur during "fight or flight" situations. One important distinction between muscle and liver is that muscle does not have a transporter to move glucose into the blood. That is, liver provides glucose for the entire body, except during times shortly after a meal. Muscle never releases glucose into the blood. It retains all its glycogen and glucose for high-demand situations.

Returning to Fig. 11.1, liver only releases glucose in response to hormonal signals, as shown at the top. The two hormones are glucagon and adrenaline. The function of glucagon is easy to remember, as it is only produced when the blood "glucose is gone"; that is, the blood glucose concentration is approaching hypoglycemia. The other hormone is adrenaline. Readers should know that *adrenaline* and *epinephrine* are identical hormones. The different names reflect a change in standard nomenclature, but each name has the same meaning. The older name, adrenaline, is from the Latin *renal* for kidney, plus the prefix "ad," meaning "on". Epinephrine has the equivalent origin from the Greek *nephros* for kidney, plus "epi" for at or on. The secretory organ that makes this hormone is still known as the adrenal gland and is located just above the kidney. Each hormone binds to its own specific receptor, but for simplicity, only a single receptor is shown.

Binding of either hormone to its receptor on the outside of the cellular membrane activates the *adenyl cyclase* (AC) that is very close by, but on the inside of the membrane. Its function is to produce a special nucleotide from the substrate, ATP, named *cyclic AMP* (cAMP). The function of cAMP is to bind to and stabilize free R subunits (regulatory) and prevent these from binding to and inactivating the *protein kinase* catalytic subunit, denoted by "C." As this enzyme's name implies, once the catalytic subunit of *protein kinase* is no longer inhibited, it will catalyze the phosphorylation of a protein, in this case, *phosphorylase kinase*. Readers will remember that enzymes named *kinases* always phosphorylate the substrate for which they are named. *Phosphorylase kinase*, as its name should suggest, then phosphorylates *glycogen phosphorylase*, thereby beginning the release of glucose molecules.

Here again, it is worth noting how efficient evolution has been. The α(1–4) bond connecting adjacent glucose molecules could easily have been broken by a specific *hydrolase*—an enzyme that would use water to break this bond. But nature has provided a slightly better enzyme, the *phosphorylase*, to split this bond, and thereby attach the phosphate at the carbon atom involved, producing Glc-1-P (glucose-1-P). In the liver cell, this feature is not that important, as the phosphate is subsequently hydrolyzed off to produce free glucose, which can then be released into the blood.

But in muscle cells, this feature is very important. At critical times, muscle cells need all the ATP that they can get. The α(1–4) bond connecting any glucose to the glycogen polymer has just enough energy that, when the bond is made to a phosphate (P_i), the phosphate can then, a few steps later, be transferred to ADP, thereby making an ATP. If a *theoretical glycogen hydrolase* were used in place of glycogen phosphorylase, all these potential ATPs would never be made. Evolution almost always finds the best answer.

Because we learned about enzyme cascades earlier (see Chap. 8), readers might recognize that the three enzymes at the top of Fig. 11.1 form a cascade. And the main benefit of a cascade is to rapidly amplify a signal, especially when this signal is only temporary. That is, the newly formed cAMP will be quickly degraded by a *cAMP phosphatase* (not shown), after which the general *phosphatase*, cartooned as a shark, will steadily dephosphorylate all the phospho-proteins and restore the cell to its normal status.

We should also note that a certain type of covalent modification, such as phosphorylation in this figure, may have opposing consequences. We see that of the four different enzymes in this figure that become phosphorylated, for three of them it makes them more active. But *glycogen synthase* becomes less active. And this is exactly necessary for the overall process being shown. The goal of either hormone at the top of the figure is to result in more glucose being released into the cell from glycogen. But the function of *glycogen synthase*, as its name implies, is to attach glucose molecules to the existing glycogen polymer. Thus, a single signal that activates the various kinases correctly activates the three enzymes that release glucose molecules, while also blocking the one enzyme that would consume glucose by making new glycogen. Our metabolism is amazingly well regulated.

Readers should also appreciate one additional feature shown in this figure. Covalent bonds, such as phosphorylation, are fairly stable and therefore long-lasting. But the two enzymes that begin and then end this process are regulated by the simple on/off binding of a regulatory subunit. To emphasize this visually, I have designed the *phosphatase* as a shark, because it steadily moves about inside the cell, looking for phosphorylated proteins from which it can cleave the phosphate residue. At the beginning of the process being depicted, the inhibitory *PPPxR*[2] itself becomes phosphorylated, which changes its conformation so that it now binds and inhibits the *phosphatase*, as if enclosing it in a cage, preventing it from undoing the phosphorylation cascade.

---

[2] The letter "x" in this acronym stands for a number between 1 and 16 for the different isozymes of this regulatory subunit. There are a corresponding 16 different *phosphatases*.

11 Control Your Sugar: Special Enzymes Control Glucose Metabolism

We see that the P-bond is cartooned as the key that closes the *PPPxR* cage. Naturally, all of the *phosphatases* will never be locked up, so that at all times some, maybe only a few, will still be free and active, and therefore also cleave the P-bond from the *PPPxR*. This means that some *phosphatases* will always be free and steadily release more of their own type and undo the phosphorylation cascade.

The actual binding of the *PPPxR* with the *phosphatase* is the same as the "*R*" and "*C*" subunits of the *protein kinase*. With the protein kinase, the change in its activity is also regulated by the simple on/off binding of its "*R*" subunit. When the "*R*" subunit binds to cAMP, it can no longer bind to the "*C*" subunit, thereby making the latter catalytically active.

This distinction in how the beginning and the end of this process are controlled is essential to curtailing the process to a limited amount of time. After all, our glycogen stores are important and should only be converted to free glucose when that is absolutely necessary. Readers should appreciate that while all cells can use glucose, only the brain and RBCs are highly dependent on glucose. Under most conditions, most of the body's cells can function very well using fat as an energy source.

Because the availability of glucose was not consistent in prehistoric times, our bodies also evolved various glucose transporters (GLUT 1–4)[3] for moving glucose from the blood into the cells of tissues and organs. As shown in Table 11.1, the four isozymes for this function have different affinities for glucose, and readers should compare these K_M values with the stated normal concentrations of glucose in blood, at about 4–7 mM. We see that red blood cells and the brain (GLUT 1 and 3) are served faster and more continuously because of their better affinity for glucose. Remember: brain and RBC are glucose-dependent, and so should NOT be insulin-dependent.

Table 11.1 Glucose transporters

Transporter	K_M for glucose (mM)	Tissue distribution
GLUT 1 & 3	2 - 10	*RBC, brain,* kidney, colon
GLUT 2	20 - 40	*Liver, β-cells of pancreas,* kidney, small intestine
GLUT 4 (insulin-dependent)	2 - 10	*Skeletal muscle, adipose tissue*

---

[3] Abbreviations for glucose are inconsistent. Both Glc and Glu are used, but Glu can also mean the amino acid glutamate.

Then the GLUT 2 transporter will generally be active only for an hour or so after a meal, when the intestine will briefly produce higher amounts of glucose from the carbohydrates in that meal or snack. The transporter for muscle and adipose tissue (GLUT 4) has a good affinity for glucose, but it must be activated by the release of insulin. And insulin release occurs mostly in response to high glucose levels in the blood—after a meal.

2   Other Sources of Glucose

In addition to the glucose polymers already mentioned, many food sources contain some additional disaccharides (Fig. 11.2). Any polymer with many sugar molecules is called a *polysaccharide*. A molecule with only two sugars linked is then a *disaccharide*, of which sucrose and lactose are the two sugars that are most common in our diet. Lactose is the sugar found in mammalian milk and is formed by a glucose sugar linked to galactose. Sucrose is the most common table sugar and is formed by a glucose linked to fructose. Sucrose is made by various plants, among which beets and cane are the most important commercially. Sucrose and lactose are hydrolyzed to their monosaccharides by *sucrase* and *lactase*.

As we age, many people develop an intolerance to the milk sugar lactose because they no longer produce adequate *lactase* in their intestine. If not digested in the small intestine, this disaccharide will be digested in the colon by normal bacteria, leading to gas and flatulence. Supplementary *lactase* pills are now available to help such people continue to enjoy dairy products.

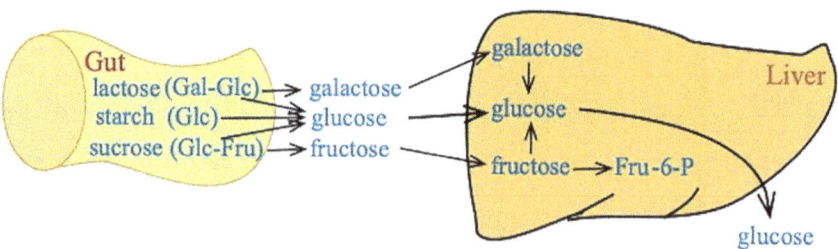

Fig. 11.2 Dietary sugars are converted to glucose

3 Glycolysis: The Pathway for Producing the Initial ATP

How do we get energy from glucose? The process begins with glucose, whose origins for our metabolism have just been described, by a sequence of 10 enzymatic reactions, normally referred to as *glycolysis*, or the *glycolytic pathway*. As shown (Fig. 11.3), glucose can come from an internal source within liver or muscle cells, or from a dietary source. To keep the glucose inside body cells and also to make it available for glycolysis, it must first be phosphorylated. This is done by *hexokinase*, originally named because glucose is a hexose, meaning that the glucose structure contains 6 carbons, as indicated at the top of Fig. 11.4.

In order to focus only on the more important control enzymes in glycolysis, Figs. 11.3 and 11.4 have two abbreviated versions to help readers see the more important components.

Humans have four isozymes for hexokinase, distributed in different tissues to optimize glucose utilization in each one (Table 11.2). In Chap. 7, it was mentioned that RBCs (red blood cells) have no mitochondria, and therefore depend solely on glycolysis to provide the necessary ATP. They have the three isozymes with the best affinity for glucose, thus enabling them to stay active for their limited lifespan in delivering the essential oxygen to all tissues, even when glucose concentrations are low.

We also see that HK I (brain and RBCs) as well as HK III (RBCs) have the lowest K_Ms (highest affinity) for glucose, guaranteeing that they will function even during hypoglycemia (low blood sugar). Readers should be aware that the stated concentrations of glucose in blood are always much higher than in the different tissues. While blood continues to gain new glucose, either from the liver's glycogen or from meals, the tissues also continuously oxidize their glucose in the glycolytic pathway, as they need ATP around the clock. Skeletal muscles are intermittently at rest, except for the heart and diaphragm, which never stop. But the brain and internal organs function fairly steadily, even while we are asleep.

In significant contrast is isozyme IV, in the liver, which has the highest K_M for glucose at 7.6 mM. The liver will not use much glucose, except for the one to two hours after a meal has been digested. The liver itself does not need glucose for its normal energy demands, as these can be supplied by fats. The liver's main use for glucose is in the synthesis of glycogen, which should occur only when glucose levels are high, so that all other tissues continue to have adequate glucose for their energy needs.

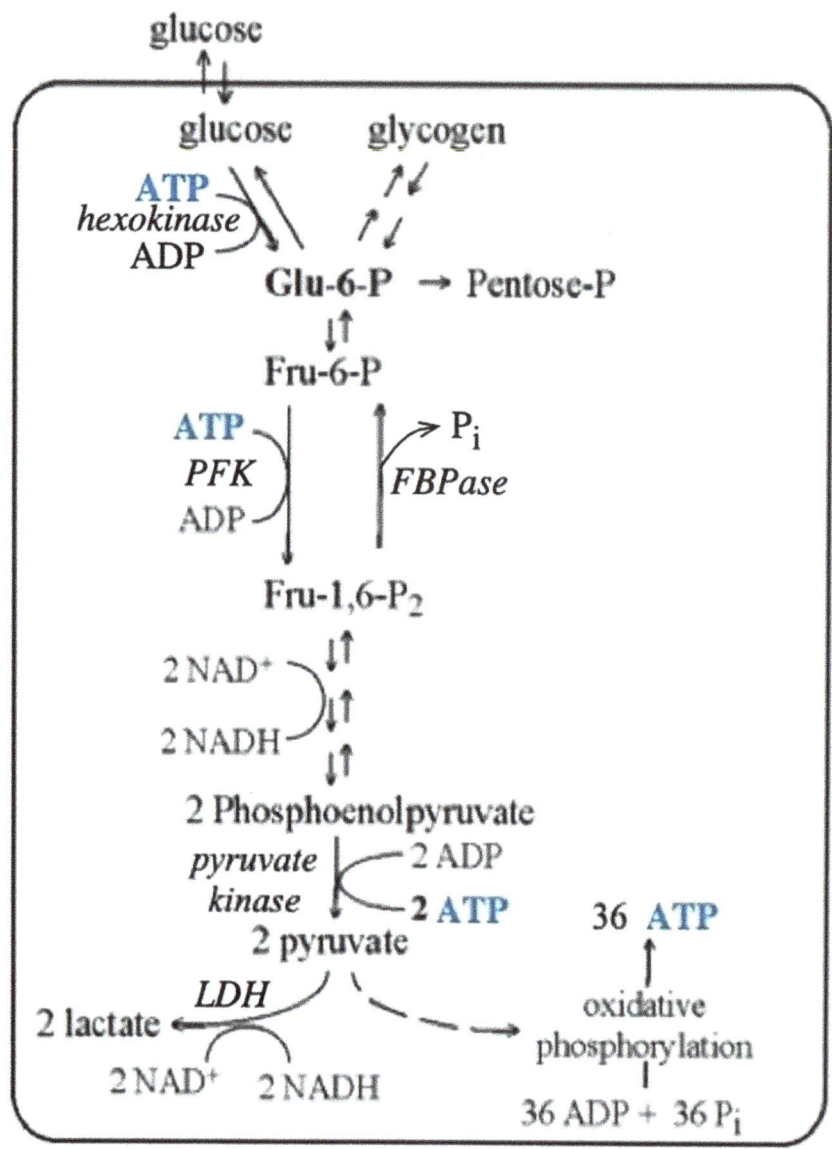

Fig. 11.3 Abbreviated diagram for glycolysis. This emphasizes ATP production, and the main allosteric enzymes: PFK phosphofructokinase, FBPase fructose-bis-phosphatase, LDH lactate dehydrogenase

The last column in Table 11.2 shows that the four isozymes vary as to whether and what type of cooperativity they show for glucose. How can four almost identical proteins vary in this attribute? The answer is shown in the second column, where we see that HK 4 has a mass of 52 kDa, while the other

11 Control Your Sugar: Special Enzymes Control Glucose Metabolism

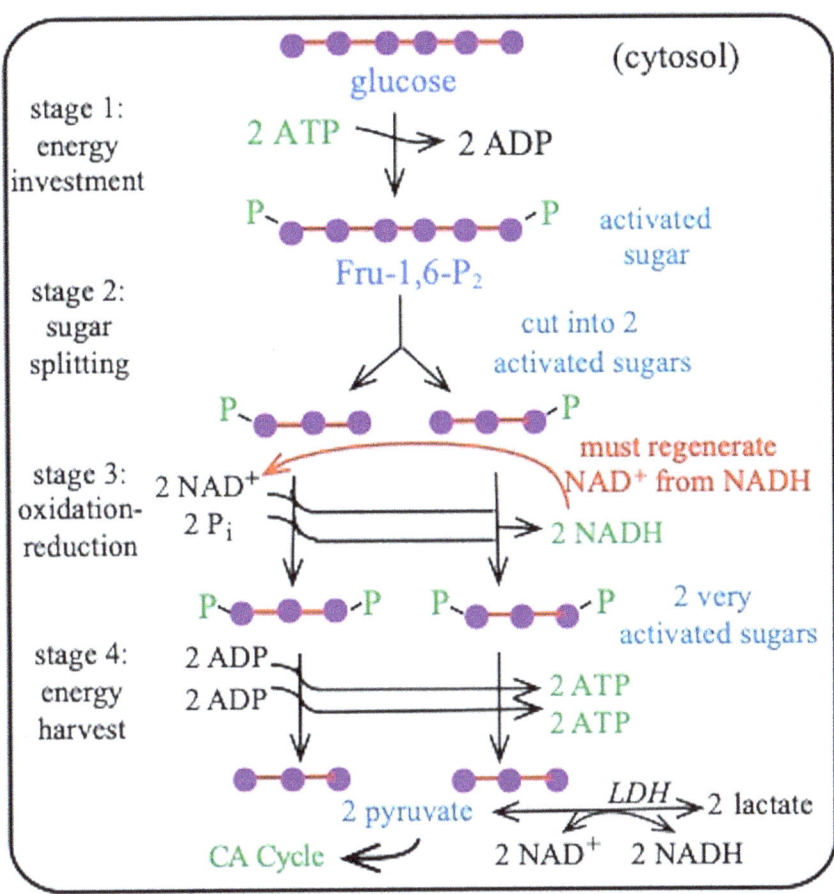

Fig. 11.4 Abbreviated diagram for glycolysis. The purple circles are the carbon atoms of glucose. CA citric acid

Table 11.2 Kinetics of human hexokinases

Isozyme	Size (kDa)	Principal tissues	k_{cat} (s^{-1})	$K_M^{Glucose}$ (µM)	K_M^{ATP} (mM)	Cooperativity for glucose
HK I	100	Brain, muscle, RBCs	100	61	1.2	No
HK II	100	Muscle, RBCs	318	340	1.0	Negative
HK III	100	RBCs	60	38	3.0	Positive
HK IV	52	Liver, pancreas	62	7600	0.2	Positive

three isozymes are twice as big. Based on their amino acid sequence alignments, it has become evident that the first three hexokinases arose by gene duplication of the smaller HK IV. Normally, when a gene is duplicated, the two genes then remain separate, and the proteins that they code for remain independent and can usually occur at different concentrations in different tissues.

However, with the hexokinases, the duplicated gene remained immediately adjacent to its ancestor, with perhaps a few extra codons between the reading frames. These extra codons could then specify just enough amino acids to form a loop between the two protein sections, each of which would have some version of the original catalytic site.

HK I has no cooperativity, suggesting that each catalytic center remained constant in its affinity for glucose. HK II has negative cooperativity. This feature normally reflects the additive sum of two separate catalytic centers that have very different K_Ms, but each center actually has no cooperativity itself. Then, as the activity at the site that binds at lower glucose begins to plateau, the second site with weaker affinity becomes active as glucose levels become high enough.

Because the two attached catalytic centers can interact with each other, then substrate inhibition is possible if the second center now binds glucose but is unable to phosphorylate it, and this somehow interferes with the other catalytic center. This inhibition would not be harmful because it only occurs at high glucose concentrations of about 1 mM or higher inside the cell. Here again, we see how evolution can exploit fortuitous alterations in a single starting gene to make four variants, and then this would naturally evolve into differential expressions of these genes as a function of the needs of the tissue in which their properties are most useful.

Our discussion of glycolysis is summarized in Fig. 11.4, which emphasizes how a 6-carbon glucose, at the beginning, is converted to two 3-carbon pyruvates at the end, as well as where ATP is used and where ATP is produced. Pyruvate, the normal end-product of glycolysis, is then used by the enzymes of the Citric Acid Cycle (CAC) to generate both NADH and $FADH_2$, which in turn provide the protons for the *ATP synthase*, where most of the cell's ATP is produced.

However, as already stated, the enzymes of glycolysis have high catalytic rates, and when the regulatory enzymes of this pathway are activated, they can produce pyruvate more rapidly than it can be consumed by the CAC. But to continue glycolysis at this higher rate, NAD^+ (stage 3) must be regenerated, so that this stage can continue to make 1, 3-bisphosphoglycerate (the 3-carbon compound with 2 phosphates). This is accomplished by the enzyme at the bottom of this figure: LDH (*lactate dehydrogenase*). This production of lactic

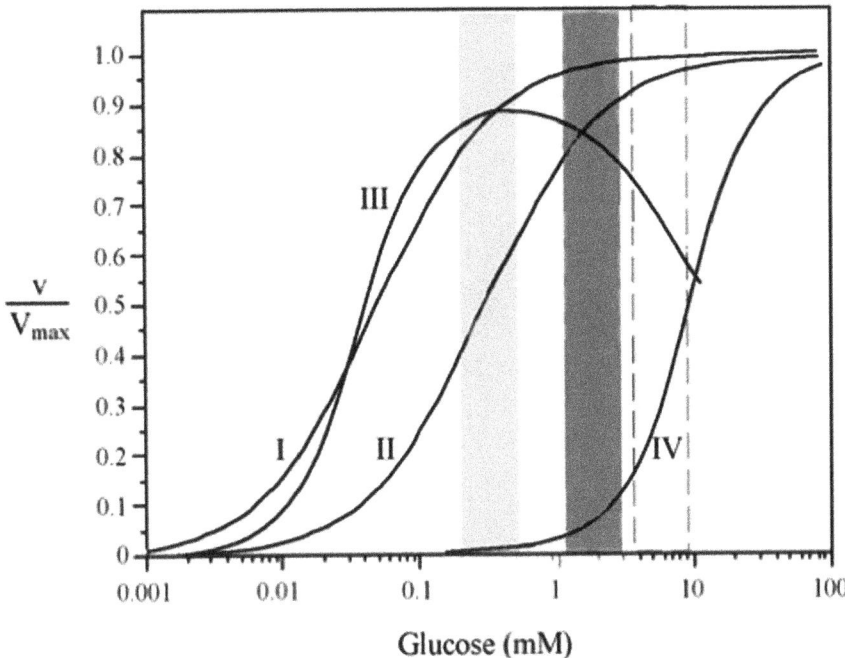

Fig. 11.5 The four human hexokinase isozyme activity curves compared with blood glucose concentrations and where the isozymes function: light shading = low (brain and other tissues), dark shading = moderate (muscle), dashed box = similar to blood (liver)

acid produces the stiffness and soreness after intense muscle exercise. After we are again at rest, LDH will reverse the reaction, reforming the pyruvate, which can now be used in the CAC.

That is also why a body develops rigor mortis (Latin: stiffness of death). When the heart stops, muscle cells do not die immediately. But without a continued supply of oxygen, the cells switch to amplified glycolysis to get energy, until their glycogen supply is totally consumed after a few hours. This leads to an accumulation of lactic acid, which is then metabolized more slowly.

A final summary and comparison for the four isozymes is shown in a different type of plot (Fig. 11.5). This has a log scale on the x-axis for glucose concentration, and this makes it more difficult for people new to such plots to discern cooperativity. I will simply affirm that HK I has no cooperativity for glucose, consistent with the data in Table 11.2. The slopes for HK III and HK IV are steeper, which indicates positive cooperativity. Positive cooperativity is helpful when glucose levels are low if the tissues with these enzymes can get energy from fats. At higher glucose concentrations, these tissues can then switch to relying on glucose and help to lower the blood glucose levels. Remember, we must avoid extensive hyperglycemia, and we are always balanced by regulatory features to keep our blood glucose neither too high nor too low. Nobody has yet defined the values for blood glucose between 4 and 7 mM as a Goldilocks zone.

One last feature is evident for HK III (Fig. 11.5). As the glucose concentration approaches 0.5 mM, which is not very high, this isozyme exhibits substrate inhibition. That is, the activity curve starts to go downward with higher glucose. This is a fairly unusual feature. The benefit of this feature has not been defined.

4   Key Regulatory Enzymes in Glycolysis

After glycolysis is initiated by hexokinase, leading to the formation of Fructose-6-P, the pathway then continues via catalysis by *phosphofructokinase* (abbreviated as PFK in figures).

Phosphofructokinase (PFK)

Fru-6-P + ATP → Fru-1, 6-P_2 + ADP
Significance: Master control point of glycolysis
Inhibitors: ATP, low pH, NADH, citrate
Activators: AMP
Hormonally-regulated activator: F-2,6-P_2

Here again, evolution has introduced some interesting features. The first is that one substrate, F-6-P, only binds when two subunits are appropriately aligned to form a complete binding site for this compound (Fig. 11.6). A similar feature is evident for the regulatory site. It has been mentioned several times that proteins such as this enzyme never exist statically in a single conformation but can briefly sample slightly different conformations. If an appropriate ligand is present and can bind at the newly formed site, that conformation becomes more stable.

Therefore, at increasing concentrations of the key substrate F-6-P, this enzyme becomes more stable in the active R conformation. In a similar fashion, the two regulatory ligands compete for the same binding site. The smaller ligand and activator, AMP, stabilizes a slight closing of the two subunits forming this site, and this aligns the binding pocket for F-6-P.

In contrast, ATP—a normal substrate for this kinase—is also an inhibitor, but at a different site. Being larger than AMP, ATP favors the structure with the two subunits slightly more apart, and this disrupts the binding site for

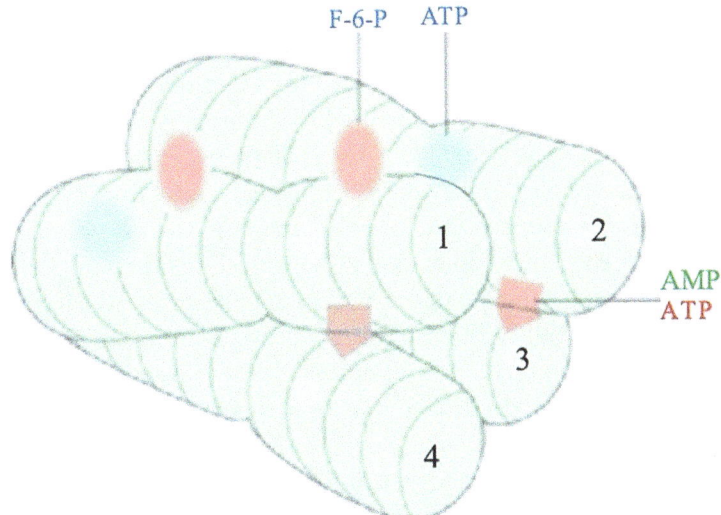

Fig. 11.6 Phosphofructokinase exists as a tetramer, with each subunit having two domains. The binding site for F-6-P (fructose-6-P) is between subunits, as is the regulatory site. AMP is an activator and ATP at high concentrations becomes an inhibitor

F-6-P. Readers might be able to deduce how these unusual and opposing functions are possible for a single compound. The answer is logical: ATP, as a substrate, has a better affinity at the catalytic site and a poorer affinity at the regulatory site. Then, if ATP concentrations are low, this substrate will still be able to bind at the catalytic site to further glycolysis and lead to the needed production of more ATP.

But, if ATP concentrations are high, the liver or muscle cells do not need to make more ATP. Instead, the glucose molecules can be used to make glycogen, which is stored for use in the future. This is a remarkable example of finely formed regulatory sites with appropriate binding constants to efficiently tune the rate of glycolysis to be consistent with the energy demands of the cell.

Pyruvate Kinase

P-enol-pyruvate + ADP → pyruvate + ATP
Significance: Produces ATP. Irreversible step.
Inhibitors: NADH, ATP
Activators: Glu-6-P, F-1,6-P_2, acetyl-CoA

If readers look at the reaction displayed above, they should see that *pyruvate kinase* is incorrectly named. After all, pyruvate kinase should phosphorylate the substrate pyruvate, which is actually its product. Instead, the enzyme is named for the back reaction, which almost never happens inside cells. This curious nomenclature occurred before scientists had fully determined all the details of glycolysis. It had been observed that P-enol-pyruvate was likely to be an intermediate. Being logical, a scientist then decided to look for an enzyme that could phosphorylate pyruvate, using extracts from cells that were likely to have such an enzyme. To observe formation of the expected product, P-enol-pyruvate, the experiments used ATP, with the phosphate group being radioactive. At various time points, samples were taken from the test tube and centrifuged to remove the proteins. The remaining fluid was then analyzed by separating components with electrophoresis (see Fig. 9.14) and looking for radioactive P-enol-pyruvate. This was successful, and the enzyme was named accordingly.

At the time these experiments were done, researchers had very little understanding about what constituted normal rates or affinities for enzymes, as described in Chap. 3. The back reaction would proceed very slowly, if at all, under cellular conditions, and in fact is almost never observed. But once the enzyme was named, that name has persisted to this day.

Lactate Dehydrogenase

Significance: regenerates NAD^+ during anaerobic glycolysis; regenerates pyruvate from lactate; freely reversible.

Lactate dehydrogenase (LDH) is not really a normal component of glycolysis and becomes involved only when glycolysis is *amplified*.

This last description is used only for muscle when strenuous muscle action is required. For such occurrences, the major pathway for making ATP, *ATP synthase*, cannot work fast enough to provide the ATP for maximum muscle activity. As has been mentioned above, the enzymes for glycolysis are above average in their concentration in muscle, and even with moderate activity rates, their production of ATP is much higher when the key enzymes are activated, leading to the glycolytic pathway being amplified.

However, the cofactor NAD^+ is an important substrate in the middle of the pathway (see Fig. 11.4), and picks up electrons to become NADH. These electrons, plus the accompanying protons, are then transferred onto oxygen by the ATP synthase, forming water as a by-product, and this also regenerates the NAD^+ so that glycolysis can continue.

But, when this pathway is amplified, an extra process is needed to regenerate NAD^+ more quickly. This is done by LDH (Fig. 11.7), which reduces

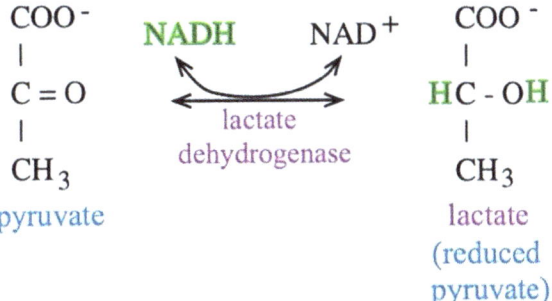

Fig. 11.7 Lactate dehydrogenase reaction

pyruvate to form lactic acid, or lactate at cellular pH, as shown in the reaction above.

Let us also briefly review nomenclature. Compounds such as pyruvate and lactate have an ionized carboxyl group at neutral pH, which is normal for cells, but are called acids at lower pH when the carboxyl group becomes protonated. They are now acids because they can transfer that proton and electron to some other chemical compound, thereby reducing it. This proton is bound weakly, so that even at neutral pH, if there is enough lactate or pyruvate, these molecules will bind available protons and thus form more of the acidic species.

A brief summary of the Citric Acid Cycle[4] is shown in Fig. 11.8, which also shows that a neural signal can lead to the release of calcium, which then activates three key enzymes to expand the rates for producing the final product ATP.

Let us now summarize how glucose, and its derivatives, are processed to produce the desired end product: ATP. Molecules in green are high-energy compounds. Metabolic energy is made in three separate processes:

1. **Glycolysis:** 1 glucose (6C) → 2 pyruvate (3C) + 1 NADH + 2 ATP
2. **Citric acid cycle:**
   1 pyruvate → 1 acetyl~CoA + CO_2 + 1 NADH
   　　　　↓ + OAA (oxaloacetate)
   　　　citrate → OAA + 2 CO_2 + 1 GTP + 3 NADH + 1 $FADH_2$
3. **Oxidative phosphorylation**
   　　2 cytosolic NADH → 2 NAD^+ + 4 ATP + 2 H_2O

---

[4] This cycle is also named the Tricarboxylic Acid Cycle (citric acid has 3 carboxyl groups) or the Krebs Cycle, for Hans Krebs who first defined it as a circular process.

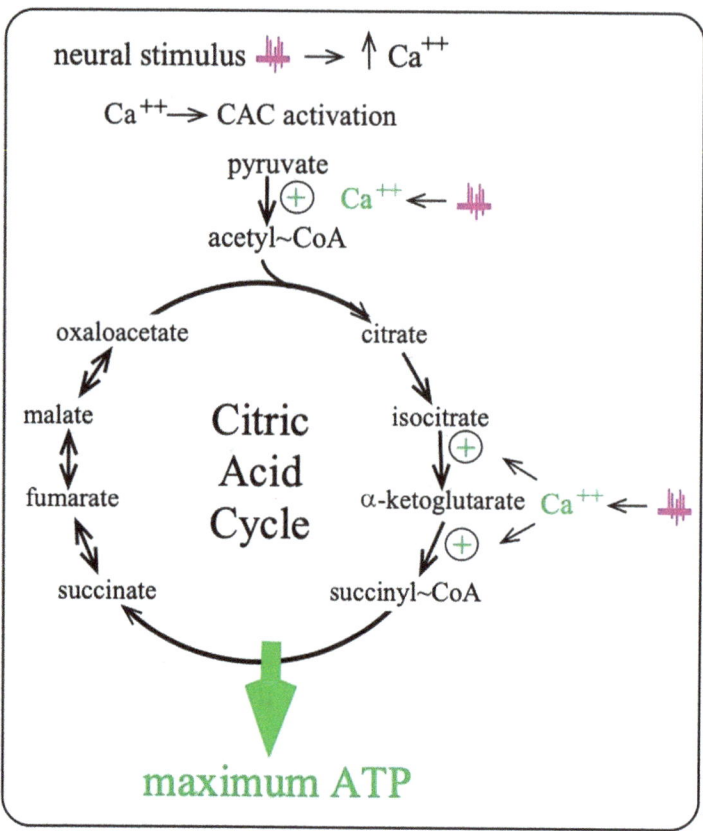

Fig. 11.8 A neural stimulus produces calcium and activates the citric acid cycle

$$2 \text{ mitochon. } FADH_2 \rightarrow 2 \text{ FAD} + 4 \text{ ATP} + 2 \text{ H}_2\text{O}$$
$$3 \text{ mitochon. } NADH \rightarrow 3 \text{ NAD}^+ + 9 \text{ ATP} + 3 \text{ H}_2\text{O}$$

In step 2 above, for the CAC, where it shows citrate → OAA, etc., that means the process goes all the way around the cycle to finally reproduce OAA.

5 Starvation

In the modern world, most people never starve, but some of us occasionally diet or fast. But in the stone age, involuntary fasting and starvation were quite frequent, and so we naturally adapted to handle these stressful situations. Figure 11.9 depicts the time course after a last meal for the body's available energy sources for survival, even for many weeks.

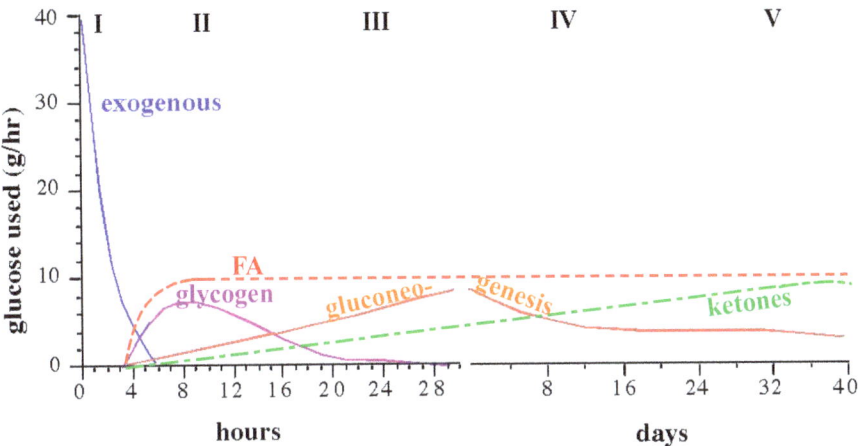

Fig. 11.9 Energy sources during fasting or starvation. Notice that the x-axis has a break, going from hours to days. FA = fatty acids

In stage I, we see that calories (mostly from carbohydrates) are gone within about 4 h. Starting at about 4 h, our natural resources, stored fat and stored glycogen, existing predominantly in the liver and muscle, are limited. Liver glycogen will be exhausted in about 20 h. By this time, *gluconeogenesis* has begun. This is the process of converting available precursors, mostly amino acids, to glucose. As already described, maintaining some glucose in the blood is almost essential for the brain and RBCs.

Acetone is formed when acetoacetate loses its carboxyl group, and because acetone is quite volatile, it will be detectable in the breath of people who are starving. If starvation continues for more than one week, *gluconeogenesis* begins to decline.

By the end of Stage II, when the liver's glycogen is almost depleted, the body begins to make ketones (Fig. 11.10) as a precursor for forming acetyl-CoA, which can then enter the CAC (Fig. 11.8). Most tissues can use ketones (see Fig. 11.10), while the brain and RBCs use the lower levels of glucose still being made by the liver, mostly by degrading muscle protein and using glycerol from fats.

It never stops completely because this process supplies the needed glucose for the brain and RBCs. To see the time courses for our energy stores, see Fig. 11.11. As already mentioned, glycogen is gone after only one day. In a normal adult, the fat deposits are moderate to slim, so the fat stores become almost depleted by 6 weeks.

Muscle continues to be converted to amino acids, which can make glucose, but this then also shrinks our muscles and makes us weaker and less able for

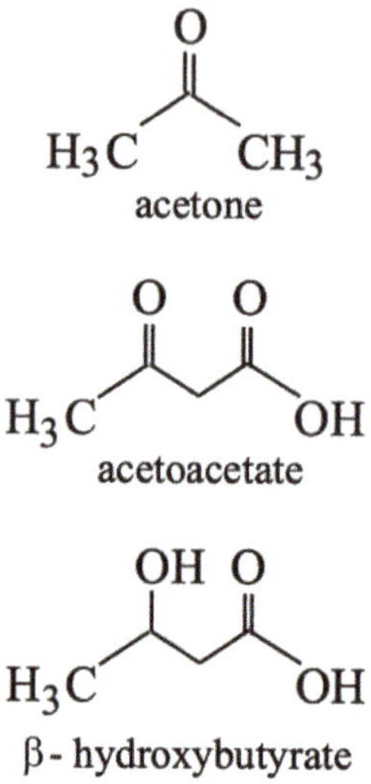

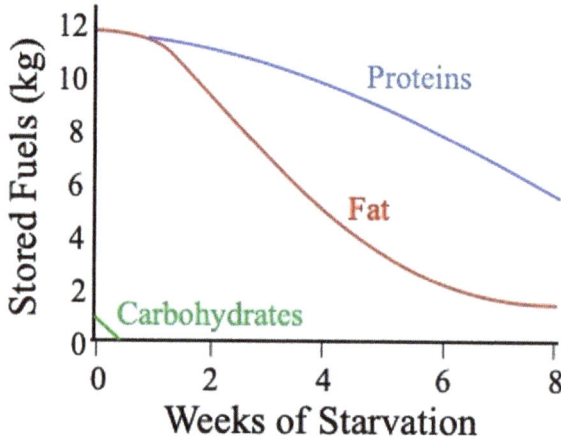

Fig. 11.10 Ketones

Fig. 11.11 Fuel sources during long-term starvation

the activity needed to hunt or gather food, at least in prehistoric times, when these processes evolved.

While fats will not be treated in any detail here, let us briefly look at the most common fat molecule—a triglyceride (Fig. 11.12). These molecules are named for having 3 fatty acids attached to the 3 carbons of glycerol, outlined in blue. Such fatty acid chains almost always contain an even number of carbons because they are synthesized by the addition of acetyl units (two carbons) to an expanding fatty acid chain. These fatty acids can then be recycled

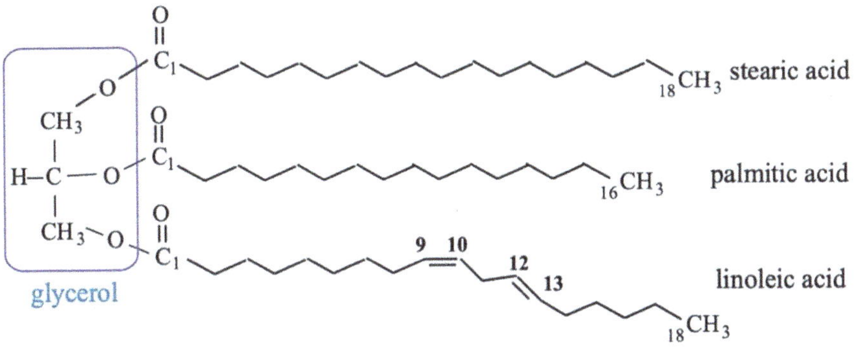

11 Control Your Sugar: Special Enzymes Control Glucose Metabolism

Fig. 11.12 A typical triglyceride containing three fatty acids. Only the first and last carbons are shown, but each high and low point represents a carbon atom. Linoleic acid also contains two double bonds

Table 11.3 Changes during starvation

	Healthy	After starving
Body weight	65 (143 lbs.)	48.8 (107 lbs.,)
Protein	11.5	8.5
Fat	9.0	2.5
Carbohydrates	0.5	0.3
Water:		
Extracellular	15	15
Intracellular	25	19
Minerals	4.0	3.5

for energy production and are converted back to 2-carbon acetyl-CoA, which can then transfer the 2-carbon group into the CAC (Fig. 11.7).

Because it contains 3 carbons, only the glycerol backbone from these triglycerides can be converted back into a 6-carbon glucose molecule. If we count the carbons in the triglyceride above, we see that it has 55 carbon atoms. Only 3 of those, almost 6% of the total, can be converted to glucose. An overview of changes in an adult, who has lost 25% of body weight during starvation, is summarized in Table 11.3. This person has lost about 16 kg in body weight. Six kilograms come from the loss of water, and 6.5 kg come from the loss of body fat. The other 3 kg are from muscle protein.

To summarize: Metabolic energy is made in three separate processes:

1. **Glycolysis:** 2 ATP + 2 pyruvates
2. **Citric acid cycle:**
    2 pyruvates → 2 NADH + 2 GTP[5] + 6 NADH + 2 FADH$_2$

---

[5] GTP and ATP have equal energy values.

3. **Oxidative phosphorylation:**

    2 cytosolic NADH → 8 ATP
    2 mitochon. $FADH_2$ → 8 ATP
    3 mitochon. NADH → 18 ATP

Since 2 pyruvates are obtained from glucose, with all three processes we obtain 38 ATPs. With glycolysis alone, we get 2 ATP. But amplified glycolysis proceeds at a much faster rate and is therefore better for those times when maximum muscle exertion is required.

Resources

Glucose metabolism: https://diabetesjournals.org/spectrum/article/17/3/183/1994/Glucose-Metabolism-and-Regulation-Beyond-Insulin

Citric acid cycle: https://simple.wikipedia.org/wiki/Krebs_cycle

Phosphofructokinase: https://bio.libretexts.org/Bookshelves/Biochemistry/Fundamentals_of_Biochemistry_(Jakubowski_and_Flatt)/02%3A_Unit_II-_Bioenergetics_and_Metabolism/15%3A_Glucose_Glycogen_and_Their_Metabolic_Regulation/15.04%3A_Regulation_of_Glycolysis

Pyruvate kinase: https://pmc.ncbi.nlm.nih.gov/articles/PMC6739817/

Metabolic changes with starvation: https://en.wikipedia.org/wiki/Starvation_response

12

Enemies Within? Our Own Enzymes Assist Viruses in Infecting Us

Abstract Description of how small viruses are, and how infectious they can be. Viruses have one or more special surface proteins with which they attach to human cells and gain entry. HIV, for example, can bind to three different cell surface proteins on human T cells. An explanation that viruses arose from incompletely synthesized bacteria, which occasionally survived as small sacs containing some RNA that coded for the necessary proteins to make new viruses. Viruses are duplicated with the aid of human enzymes. Being generally inside human cells, it is difficult to develop medications that will not also harm the human cell. Common flu viruses have two surface proteins: H (hemagglutinin) and N (neuraminidase), which are recognized by our antibodies. Flu viruses have a very sloppy polymerase, so that viral proteins mutate frequently; there are now 18 types of H proteins and 11 types of N proteins, so that more than 130 variants of these two combinations have been observed. That is why we need new flu vaccines about every 2 years. The virus causing HIV-AIDS infects our T lymphocyte cells, and therefore people do not die from the virus itself, but from various infections that occur after the immune system ceases to function. A subset of people does not have a complete CCR5 receptor, making them immune to viral infection. It is thought that this mutation emerged during the epidemic called the black death plague in the fourteenth century. The virus spreads by having its RNA reverse transcribed to DNA by the viral *reverse transcriptase*, and then becoming inserted into human DNA, where it will remain. The SARS Covid virus causes **s**evere **a**cute **r**espiratory **s**ymptoms and has a surface spike protein that binds to the ACE2 enzyme on epithelial cells. Therefore, it can reach and infect every cell in the body. Modeling of this spike protein has identified numerous "hot spots" where future mutations could occur and make this protein no longer recognizable by current vaccine antibodies. Human deaths from Covid exceed mortality numbers for all wars.

Keywords Viruses • T cell • Membrane receptors • Hemagglutinin • Neuraminidase • Vaccine • RNA polymerase • HIV • AIDS • Immune variants • Reverse transcriptase • Spike protein • ACE2 enzyme • Hot spots

In the world of tiny things, nothing is smaller or more devastating than viruses that we cannot see with the naked eye but that hold the potential to kill millions and bring the world economy to a halt. Viruses, like zombies, are not living, so they can't be killed by normal medicines. Many different infectious agents, such as bacteria, worms, and fungi, are living organisms and can be killed. Bacteria, for instance, can be killed by antibiotics. Living organisms have the ability to acquire nutrients from their surroundings so as to grow into their full form and become able to produce descendants so as to maintain their population, or better—increase their population.

Viruses are unique in that they cannot physically move by themselves and are not able to use nutrients to make new viruses by themselves. However, they are also very durable: a viral particle can be frozen or kept in a vacuum without damage. Only high heat or high acidity will normally destroy viruses before they enter a human body.

1   How Do Viruses Invade Us?

Viruses are remarkably efficient infectious agents. There are about 900 different viruses that infect at least one or more animal species. For the virus to cause an infection, it must first enter the cells of the host organism. This infection process depends on at least two separate features: the host cell must have at least one specific cell-surface protein (one that sticks up or out of the cell; see Fig. 12.1), and the virus must have at least one viral surface protein (see Fig. 12.2) that uniquely recognizes this host cell protein and binds to it (see Fig. 12.1).

Essentially, viruses are like zombies: not truly alive, and dependent on living cells for their maintenance and reproduction. The virus simply contains the genetic instructions for making more viruses, and because it can invade

12 Enemies Within? Our Own Enzymes Assist Viruses in Infecting Us

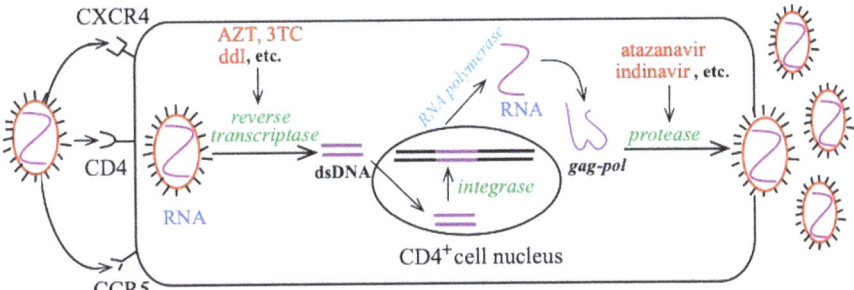

Fig. 12.1 A T-cell with various co-receptors on the cell membrane that enable HIV to bind to the cells, and this leads to fusion of the viral and cell membranes to enable the virus to enter the human cell. Inhibitor drugs are shown in red. Viral enzymes are green; human *polymerase* is blue. Compared to the cell, the virus would be much smaller than drawn

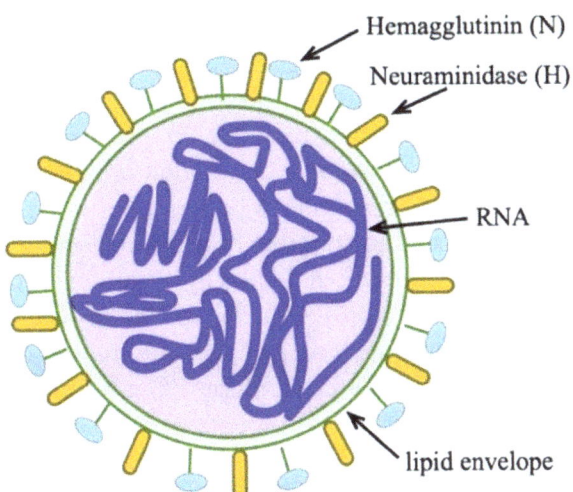

Fig. 12.2 The common influenza virus. Extending from the viral envelope are two important outer proteins, N and H, which are recognized by human antibodies. The neuraminidase enzyme cleaves the sugar link of neuraminic acid on the cell to be invaded

living cells, it can then be produced in very large numbers to be spread by blood to other cells inside our body. If the infected person coughs or sneezes, the virus can also spread by air or by water to other organisms. Because viral infection normally results in the infected cells dying, this is quite harmful to the infected organism, such as humans.

2 Why Are There Viruses? What Are They Good For?

Viruses are not necessary for any biological purpose. Life for all organisms would be better without them. In contrast, while some bacteria can cause serious diseases, most bacteria are harmless to humans, and many hundreds of bacterial species live on our skin or within the intestines of humans and may provide a biologically important function. We benefit from their functions in maintaining proper digestion and health.

All viruses require host cells in which to replicate, utilizing the host cell's polymerases and protein synthesis apparatus to express the viral genes. Viruses also depend upon the host cell's transcription and DNA replication enzymes. Cellular dependence is not a property unique to viruses; some bacteria (e.g. *rickettsia* and *chlamydia*) can grow only within a host cell. Since the critical steps in viral replication rely on the essential host cell gene expression machinery, it is difficult to develop antiviral drugs that target these critical steps without impairing host viability. Thus, vaccines are the best solution to limiting viral diseases. Viruses originated as "unused parts" from the process whereby bacteria undergo cell duplication. This duplication process is normally sufficiently effective to produce thousands of bacteria. But it is not perfect, and sometimes a bacterial cell division may not be completed, leaving parts behind from an unfinished new bacterial cell. Normally, the unfinished parts are simply degraded, either by some nearby living bacteria or by our human lymphocytes (if this occurs inside our bodies). Lymphocytes are the protective cells of our immune system and include macrophages (Greek for "big eaters") that devour foreign cells or viruses.

Occasionally, some of these spare parts, if they contain fragments of the original bacterial DNA or messenger RNA, can become encapsulated in a vesicle (a small sac formed by membranes; Fig. 12.3). Being enclosed by this vesicle allows viruses to resist being degraded. Even less frequently, such a vesicle may acquire a surface protein, such as the spike protein, by which it can now bind to a surface protein on a bacterial cell or the ACE2 enzyme on a human cell (Fig. 12.4).

When this happens, a meaningless spare part becomes an invading infectious agent. Such new viral particles will be "successful" for the virus if the fragments of genetic information (the RNA in Fig. 12.2) just happen to contain the information for reproducing the viral molecules into more identical copies of this virus. Therefore, the only purpose of a virus is to infect a cell,

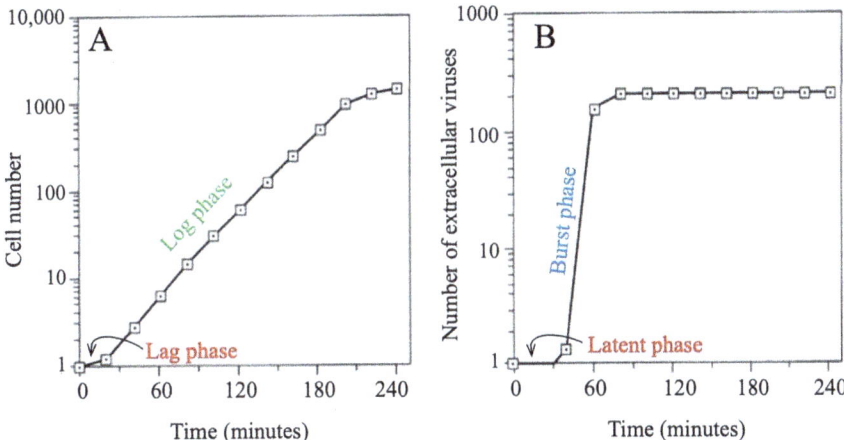

Fig. 12.3 Growth rates for bacteria (a) and viruses (b), measured in cell culture. Note that the y-axes have a logarithmic scale

where it commandeers the established cellular machinery that cells use for their own functions to make hundreds of copies of the invading viral particle.

Viruses are *small*. Unlike most bacteria, viruses cannot be seen with a light microscope, and they can pass through sterilizing filters used to remove bacteria (see Fig. 2.2 in Chap. 2). Note that some small bacteria (e.g. mycoplasma) can also pass through sterilizing filters. Viruses are a nucleic acid wearing a protein coat. Viral genomes contain a small number of genes, from 3 to about 100, depending upon the viral species.

Although the smallest bacterial genome currently known contains ~250 genes, most bacteria have at least a few thousand genes.

Unique to viruses is their biochemical composition. *Most viruses contain only a single type of nucleic acid, either DNA or RNA* (all cells contain both DNA and RNA). The remainder of a virus is protein or glycoprotein. Viruses do not possess (their own) lipids, glycolipids, sugars, polysaccharides, and metabolites such as ATP, amino acids, etc.

Also unique to viruses is their way of replication. If a single bacterium is introduced into culture medium and one plots on a semi-log scale the number of bacteria as a function of time, the data will fall on a straight line (Fig. 12.3a). This result shows that bacterial growth is exponential, i.e. one cell gives two, two cells give four, four give eight, and so on. In other words, the bacteria grow by dividing. Eventually, however, there will be no further increase in cell number and, if cells at this stage are placed in fresh media, most will no longer grow.

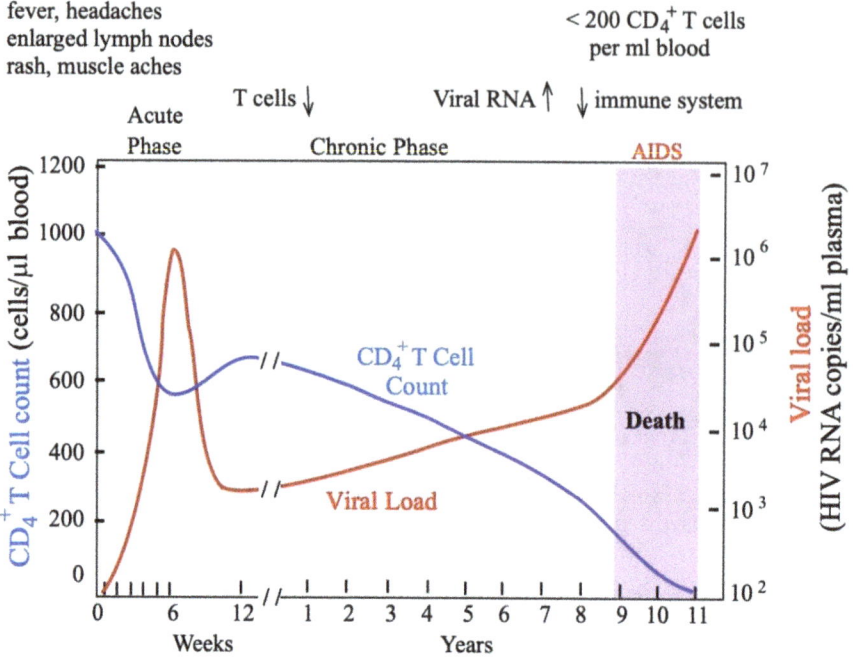

Fig. 12.4 Progression of AIDS after infection with HIV. Infection leads to depletion of CD4+ T cells, and this reduces immune competency and leads to death after infection by other pathogens. Note that the axis for viral load is logarithmic. Note that the time axis has separate sections with different scales. (Adapted from https://en.wikipedia.org/wiki/File:Hiv-timecourse_copy.svg)

In contrast, if a single virus enters a cell, for some period there is no increase in virus number (the latent phase). Then there is a sudden, nearly instantaneous increase in virus number, going from zero to hundreds of viruses, and then no further increase (Fig. 12.3b). This phenomenon, referred to as *burst kinetics*, is not an exponential process. At the end of the process, all of the viruses produced can re-infect other host cells.

The reason why a plot of virus number versus time is not exponential is because, after entering a cell, a virus replicates by expressing all of its genes and making many copies of its genome. As the abundance of these proteins and genome copies reaches a critical threshold, all of the virus particles assemble spontaneously and are released from the cell at once.

3 Common Influenza Virus

Almost everyone has experienced the common flu. And most likely, this experience has occurred many times. Such flu viruses frequently originate in birds and are often passed on to pigs, which are quite abundant on Chinese farms, and then are passed on to the attending farmers. This is why many flus in the past were called the Asian flu.

The CDC (Centers for Disease Control) states that infection rates in the United States vary from 3% to 11% of the population for any flu season. This may seem surprising, considering that we just explored how amazing our immune system is, and how it protects us from infectious agents after it has had a brief exposure to them.

But viruses have evolved to be experts at constantly changing their surface proteins (see Fig. 12.2), which are what antibodies recognize. Also, there are four separate clades, or groups, of flu viruses: A, B, C, and D. For the A clade, there are 18 different hemagglutinin subtypes and 11 different neuraminidase subtypes (H1 through H18 and N1 through N11). Simple math then shows that there are potentially 198 different combinations of these two proteins, and for influenza A, more than 130 variants have been detected so far.

In Chap. 9, we saw how amazingly efficient the human DNA polymerase is because it makes very few errors. By comparison, the flu virus, similar to the HIV virus (Table 12.1), is remarkably sloppy. It makes two mutations for each newly formed copy of the virus. While this appears to still be better than in humans, we must remember that in humans almost 99% of mutations do no harm (see Chap. 9). This is also a numbers strategy for the virus. After infecting an epithelial cell, up to several hundred copies of the flu virus will be made. If only 10% of them survive, meaning that their genes have not changed enough for the virus to become nonfunctional, then those mutations that lead to an amino acid change in one of the surface proteins will make such viruses no longer recognizable by the antibodies that were good at detecting the earlier version. That is why we need new flu shots at least every other year.

Table 12.1 Mutation rates

Species	Genome size (bp)	# cells	Polymerase fidelity (errors/bp).	Mutations/ cell division	# mutations/ lifetime
HIV	9748	1	2.2×10^{-4}	2.1	2.1
E. coli	4.6×10^6	1	$\sim 1.0 \times 10^{-9}$	2.5×10^{-3}	2.5×10^{-3}
H. sapiens	3.2×10^9	$\sim 1 \times 10^{13}$	$\sim 1.0 \times 10^{-9}$	6	$\sim 5 \times 10^{15}$

The common flu virus infects mainly our epithelial cells, meaning that initially the virus largely infects the nasal and oral passages as well as the lungs. Once inside any cell, it uses the normal enzymes inside these cells, such as the RNA polymerases and the ribosomes (see Fig. 12.1), to make hundreds of copies of its own RNAs and proteins.

4   HIV-AIDS

Almost everyone is familiar with the abbreviation HIV-AIDS. But it is easy to forget what the initials stand for: Human Immunodeficiency Virus—Acquired Immunodeficiency Syndrome. Both names include *immunodeficiency* because that is the central feature of this infection.

Let us briefly examine the time course for the progression of this disorder following infection by HIV (Fig. 12.4). After entering a cell of our immune system, the virus is very quickly reproduced by the host cell enzymes. When that cell dies and releases all the newly formed viruses, the viruses will quickly infect adjacent T cells. But most of our immune cells are still healthy and now have mounted an attack on the virus particles that are accessible, in some fluid, such as blood or lymph. Our antibodies cannot detect viral particles that are still inside cells. So, by 12 weeks, the viral load is much lower, and our immune cells are also multiplying, leading to a moderate increase in their number. After about one half of a year, with the rate of infection being faster than the rate of immune cell duplication, the process enters the chronic phase, where the viral load goes slowly but steadily higher, while the number of healthy T cells declines. It is this lowering of our natural protective immune process that enables the viral load to continue to increase.

The HIV virus does not directly kill infected people. But because they no longer have adequate immune function, most HIV patients die from other infectious diseases, such as pneumonia. As shown (Fig. 12.4), the overall process is very slow before infected people die. This helps the virus because the infected person, especially in the early years, when such infected people do not feel very ill, can easily spread the virus during sexual contacts.

5   The Resistance Allele

Attachment of the virus to immune cells is via a normal receptor on these cells for *chemokines* (*chem*ical cyto*kines*: small protein molecules that are released during infection and promote an inflammatory response). The surface protein on the virus is recognized by the CCR5 receptor, and also by the CXCR4 receptors (Fig. 12.1).

A significant fraction of about 12,000 hemophiliacs who received tainted blood failed to develop AIDS. It was found that they had a shortened version of CCR5, lacking 32 amino acids, and this modified receptor either fails to reach the cell surface or is so deformed that it cannot bind HIV. Among infected individuals, none had the mutant allele, while 3% of uninfected people had the resistant allele (Table 12.2). While the difference between 0% and 3% may seem small, the fact that no infected people had the mutated resistance allele was a very important result. Being homozygous (having the same gene from each parent) for the deletion allele is protective against HIV.

No homozygotes for the mutant allele have been uncovered among Africans, Asians, or African-Americans. The mutation must have arisen sometime after humans left Africa about 65,000 years ago. Some devastating event in Europe apparently gave originally rare individuals, who harbored the mutant, a dramatic survival advantage. The catastrophic challenge may have been a major epidemic caused by an agent that, like HIV, makes use of the normal CCR5 protein to infect cells.

The virus that caused the black death plague (fourteenth century) also binds to CCR5. Individuals homozygous for the CCR5 mutant would have survived. The heterozygous genotype helps (about 20% of Caucasian-Americans), but only delays the time at which CD_4^+ T cells fall below 200/μl of blood (i.e. infected individuals live longer but carry the virus).

It is known that the HIV virus originated in Africa, among monkeys, and somehow spread to local people. It is most commonly spread by oral or genital fluids, and that is the path by which local African people may have become infected. Efforts to trace the spread of this disease to the United States have identified a potential initial person. This was an airline pilot who occasionally flew flights from San Francisco to Africa. The virus then spread quickly in the gay community because these men frequently had multiple partners.

However, that also meant that the virus was initially not deemed that serious, as it appeared to be restricted to people who were outliers in normal American culture. The disease had been known for several years before serious efforts commenced to deal with it. The reason this changed is that some of the gay men, now infected, were also blood donors. The virus therefore spread into the general population by blood transfusions at local hospitals.

It was then soon determined that the main cells being infected in our immune system were CD_4^+ T cells. These are so named because they have the CD_4 surface protein on their membrane (Fig. 12.1). Additional studies then showed that the virus has an RNA genome, which is transcribed into

Table 12.2 Resistance allele

	2 copies of standard CCR5 allele	1 standard and 1 mutant CCR5 allele	2 copies of mutant CCR5 allele
HIV-infected individuals	85%	15%	0%
Uninfected Individuals	83%	14%	3%

double-stranded DNA inside the host cell. This is done by a viral *reverse transcriptase*. The term reverse is part of this enzyme's name because its catalytic function is opposite to our normal RNA polymerases, which transcribe DNA to make RNA copies (see Fig. 9.11).

This newly made small copy of the viral double-stranded DNA is then inserted into the human T cell's own DNA by an *integrase*. Once the viral DNA is packaged inside the human DNA, then human *RNA polymerases* will routinely transcribe it into many copies of viral RNA. These RNAs are then translated into viral proteins by the host cell's ribosomes.

An unusual feature of the HIV genome is that it does not have a separate gene for each of the few proteins it needs. Some of these coding regions have no separation between them, so that the RNAs code for polyproteins, such as the *gag-pol* in Fig. 12.1. Gag-Pol is an abbreviation for two separate proteins: a glycose-amino-glycan coated coat protein (Gag), and the viral reverse transcriptase (Pol; because it is a polymerase).

The process diagrammed above (Fig. 12.4) is slow but continuous. While symptoms of lowered immune function may occur early, death does not normally occur until many years have elapsed, and almost all immune T cells have been destroyed by viral infection.

6   SARS-CoV-2 or Covid-19 Virus

This virus is again known by almost all people, but generally as an acronym. SARS represents the disease by its symptoms: **S**evere **A**cute **R**espiratory **S**yndrome. Covid has a similar meaning but stipulates the pathogen: **Co**rona **vi**rus **d**isease, where the term *corona* (from the Latin for crown) refers to the appearance of the virus, surrounded by the spike proteins (Fig. 12.5). The number 19 refers to the year it was first identified in China: December of 2019.

The number 2 identifies this strain as the second SARS variant. The first, also originating in China, occurred in 2002–2004, but the public and medical responses were sufficiently effective that the virus was quickly eradicated.

The Covid virus depends on several human enzymes for its survival (see Fig. 12.7). Initially, the virus uses its own *RNA polymerase* (#3) to make more copies of its genome, which are then used by the human ribosomes (#2) to make viral proteins. But the virus can also make DNA copies from its genomic RNA using its own *reverse transcriptase* (#4), and with its own viral integrase (#5), this Covid DNA can now be inserted into any available human chromosome. Then the human *RNA polymerase* (#1) can continue to make viral RNAs.

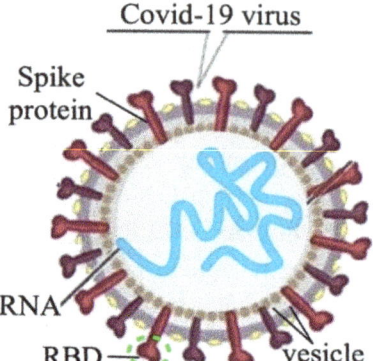

Fig. 12.5 A SARS CoV-2 virus. *RBD* Receptor binding domain

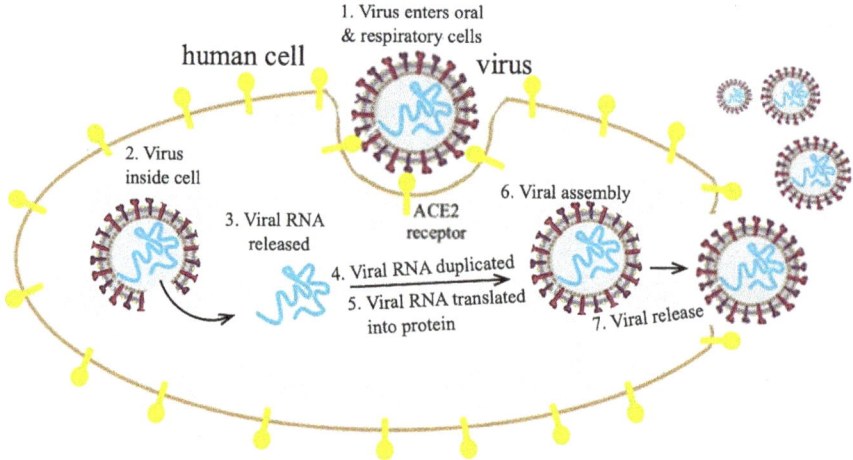

Fig. 12.6 The route by which the Covid virus enters a human cell, where many copies of the virus will be made, and released, to spread the infection to other cells or hosts. The yellow membrane proteins represent ACE2 (Angiotensin Converting Enzyme #2), found on all cells lining human blood vessels

The receptor-binding domain at the end of the spike protein (Fig. 12.5) had evolved to attach to an enzyme on epithelial cells: *angiotensin-converting enzyme 2* (Fig. 12.6). Angiotensin is a peptide hormone that causes vasoconstriction of arteries, and such constriction increases our blood pressure. The enzyme converts the hormone precursor, angiotensin I, to the active angiotensin II. All epithelial cells have this protein in their membrane, facing toward the inside of the blood vessel, where angiotensin I would flow by, after being released. Since every cell in the body is next to or very close to a capillary, the virus can enter every cell in the body.

12 Enemies Within? Our Own Enzymes Assist Viruses in Infecting Us

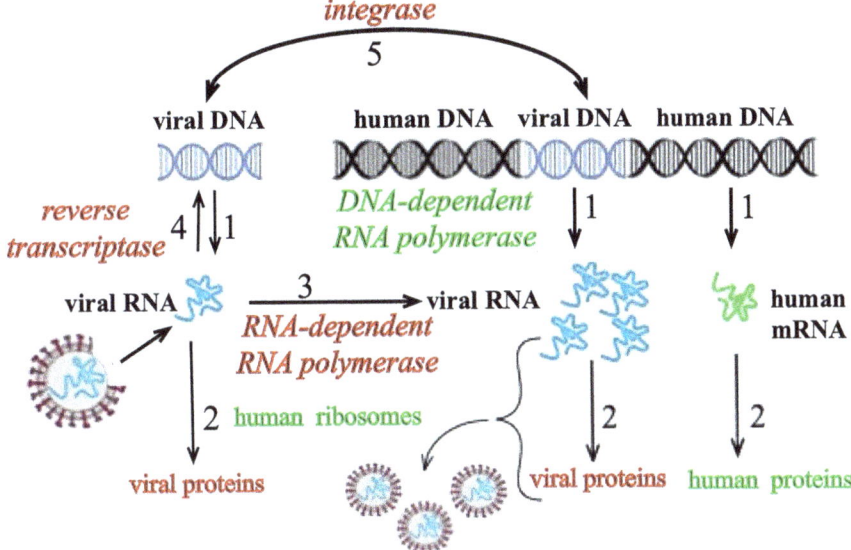

Fig. 12.7 Process for viral enzymes (red) to produce more viral RNA, or to insert new viral DNA into human DNA. Human enzymes in green

Like other viruses, for Covid-19, the most common source of entry is via the oral or nasal passages. This routinely leads to inhaled air, containing the virus particles, going to the lungs initially, leading to severe acute respiratory distress—the initial symptoms for which the disease is named. But with time, the virus may slowly spread, by the same entry mechanism, into cells in the brain or other organs. And this leads to various symptoms under the umbrella name *Long Covid*.

A close-up view of the RBD protein (Fig. 12.8) shows "hot spots" where future mutations have been predicted, and such changes would avoid the existing antibodies that are currently in the human population. Such predictions are possible based on our current knowledge about the structure of the spike protein. Therefore, unless people learn to develop better hygiene to protect themselves, the Covid virus will return in new variants for many years to come.

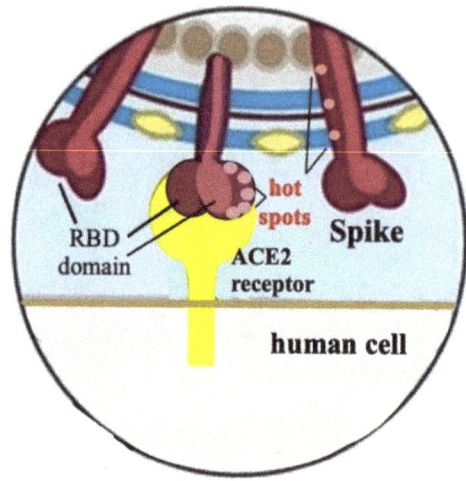

Fig. 12.8 Predicted mutation sites (hot spots) in the spike protein's RBD and stem

Table 12.3 Major causes of death in the US

Event	# of American deaths
American civil war	About 700,000
World war I	About 116,000
1918–1919 flu	About 670,000
World war II	About 272,000
Covid-19 flu	Over 1,220,000

7 How Bad Is Covid?

Published numbers may have limited accuracy for mortality rates, because a Covid death may easily be ascribed to some other cause, especially when patients had/have simultaneous other medical conditions. For the first SARS-CoV-1, mortality rates were about 10% of the infected population. For SARS-CoV-2, mortality rates have been estimated at about 7.2% of the infected population.

Another way of looking at the pandemic in the 2020s is to compare it with other events in which there was a large loss of life. We then see that pandemics are far deadlier than any of the wars (Table 12.3).

Resources

Viral infections: https://my.clevelandclinic.org/health/diseases/24473-viral-infection
Flu virus: https://pmc.ncbi.nlm.nih.gov/articles/PMC3074182/
HIV virus: https://www.cdc.gov/hiv/about/index.html
Covid virus: https://en.wikipedia.org/wiki/COVID-19

13

Traitors: Cancer Results When Regulatory Proteins Become Mutated

Abstract There are more than 200 different types of cancer. Tumor cells normally multiply rapidly. Cancers commonly require at least five distinct mutations in the same cell before it becomes a cancer. Such mutations take time, so that cancers occur more frequently as people become older. For example, Trisomy 21 (Down syndrome, and not a cancer) increases greatly as mothers approach the age of 40. In a similar manner, the likelihood that offspring will have neural defects increases with the age of fathers in a survey in Iceland. Description of the cell cycle and checkpoints in that sequence of events: cells in the G_0 stage have left the cycle and can differentiate to achieve their special function. Cells in the G_0 stage can re-enter the cell cycle for replication only in response to some specific signal to activate this process. This normally involves cyclin-dependent protein kinases (Cdks). Most cancers result from mutations to and subsequent improper function of tumor suppressors or oncogenes. Rb was the first tumor suppressor identified in children with retinoblastoma. A transcription factor known as E2F (early region 2 binding factor) acts to attract RNA polymerase to the TATA box, leading to the synthesis of proteins needed for cell division, and this is blocked by the Rb suppressor. Cdks phosphorylate Rb, making it inactive. The p53 tumor suppressor is the most important for regulating cell division. Explanation of western blots: some viruses can cause cancer.

Keywords Cancer • Hepatoma • Sarcoma • Lymphoma • Tumor • Trisomy 21 • Neural problems • Neoplasm • Dysplasia • Benign tumor • Mitosis • Interphase • Checkpoint • Metastasis • Cell cycle • Cyclin-dependent protein kinases • Cdks • Cyclin • Tumor suppressor • Oncogenes • Rb tumor suppressor • p53 tumor suppressor • Reactive oxygen species • Retinoblastoma • Western blot • Oncogenes

Most people have heard or read statements such as "It is time to cure cancer." This suggests that cancer is somehow a single type of disorder and, therefore, susceptible to being treated by a single procedure or chemical therapy. In reality, there are more than 200 different types of cancer, and they arise from a variety of initial causes. Most of us are familiar with some of the more common cancers: hepatoma (liver cancer), lymphoma (cancer of lymph nodes), and sarcoma (cancer of connective tissues). In fact, there are more than 70 types of sarcoma. Any term ending in *oma* (from the Greek for morbid growth) is a cancer. It is a morbid growth because cancer cells normally keep expanding unchecked and often lead to death.

Tumors grow rapidly (Fig. 13.1), as the number of cell doublings soon reaches a size that becomes significant. As shown in this figure, tumors become detectable by simple examination when they are about 10 mm in diameter—the size of an eraser on a pencil's end.

With our discussion of sickle cell anemia, we have an example of a "one-hit" genetic disorder, meaning that a single gene must be mutated in both chromosomes carrying that gene. This is almost always possible if some individual has inherited one mutated version from a parent, who was heterozygous, and then has the misfortune of having a chance mutation occur later in the other, normal version of that gene, making this person homozygous for the mutation or disorder.

We have a few examples of "one-hit" cancers, meaning that only one gene must be mutated to cause the cancer. But most cancers require four or five hits; that is, four or five different genes, coding for different proteins, must be mutated in the same cell before the cancer begins. Because these "hits" occur by chance and randomly, it is therefore fairly common that many years are required before all four or five genes experience the required mutation. This is why most cancers do not arise until people are in mid-life or older. While Trisomy 21 (formerly named "Down syndrome") is not a cancer, it illustrates how multiple hits are acquired with age to produce infants with this disorder—an extra copy of chromosome 21.

We see that the probability of an infant being born with this syndrome rises dramatically as the mother becomes older (Fig. 13.2). A similar result was seen with the probability of neural problems in children of older fathers (Fig. 13.3).

Let us start with some vocabulary:

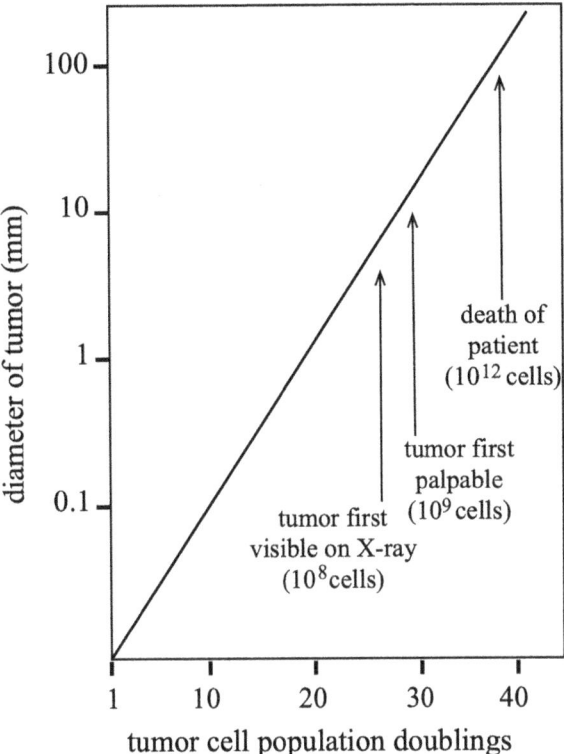

Fig. 13.1 Increase in tumor size with the number of doublings of tumor cells. (Adapted from https://www.aacr.org/wp-content/uploads/2019/11/SLIDES-FDA-AACR-ASTRO-Workshop-Day-2.pdf)

1. **Neoplasm/neoplasia**: "New growth" results after cell transformation. The name *cancer* is derived from the Latin for "crab" because of the spreading fingers of invasive growth that characterize malignant neoplasms.
2. **Dysplasia**: A combination of abnormal cellular appearance and abnormal tissue architecture in an early neoplasm.
3. **Benign tumor:** Encapsulated nodules of neoplastic tissue that do not spread and grow with a smooth, pushing edge. Example: Colon polyps, fibroids of the uterine myometrium, and endocrine adenomas due to excessive hormone production. They are named "benign" only because they have not yet metastasized, so that they remain a single lump that can be surgically removed.
4. **Malignant tumor**: Invades adjacent tissue and is considered to be cancer. Some exceptions exist: Basal cell carcinoma of the skin is histologically malignant, invades aggressively, but does not metastasize.

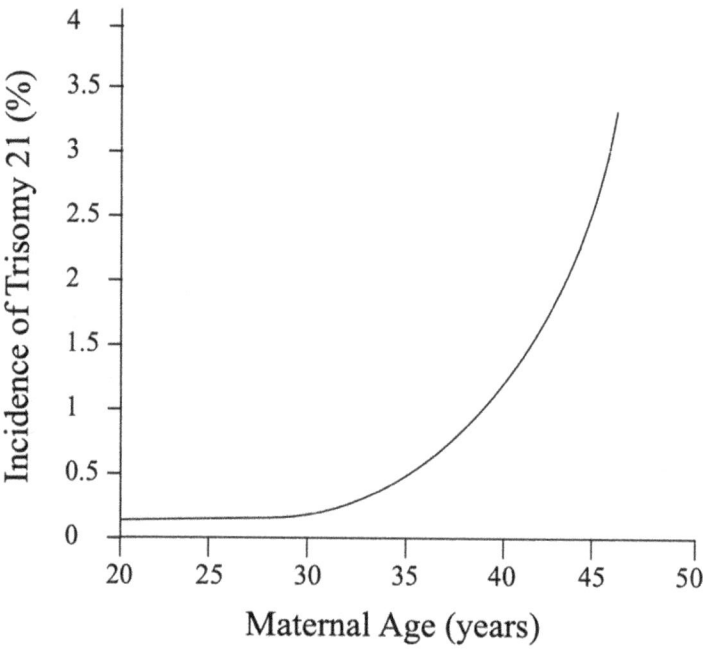

Fig. 13.2 The increase in infants with Trisomy 21 as a function of maternal age. (Adapted from Research Gate, via Creative Commons Attribution 2.0 Generic)

5. **Metastasis**: Spread of tumor cells to neighboring tissues = *invasion*. Spread to distant organs = *metastasis*.

1 The Cell Cycle and Cancer

Single-celled organisms grow and divide when essential nutrients are available, and they stop growing and dividing when nutrients are absent. Multicellular organisms, which are generally isolated from environmental fluctuations, follow a completely different set of growth rules. A single-celled zygote undergoes countless divisions. Along the way, subsets of cells differentiate morphologically and biochemically to carry out specific functions (e.g. a liver cell or nerve cell). Generally, differentiated cells stop dividing; otherwise, the organs and tissues they form would not be properly organized. Some cells continue dividing; for example, hematopoietic stem cells, which continue producing new red and white blood cells, while epithelial cells lining organs (e.g. stomach, intestine, outer body) are continually renewed. Finally, some non-dividing cells can be made to divide by specific signals, e.g. lymphocytes

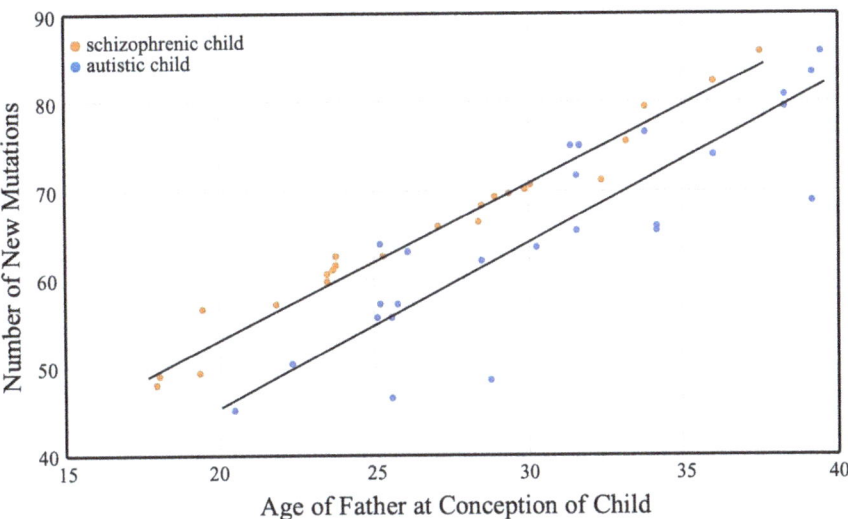

Fig. 13.3 Increased neural problems in children of older fathers. Caveat: Data are from Iceland, a small, isolated, and therefore inbred population. It is possible that this has produced some widespread defect in the male population, leading to higher mutation rates. (Adapted from: Nature (2012) 488: 471–475)

will proliferate through interaction with an appropriate foreign substance (antigen).

What determines whether cells in multicellular organisms will or will not grow and divide? These decisions are made by a specific biochemical program, the *Cell Cycle*. The Cell Cycle consists of all the events that occur from one mitotic division to the next, during which all of a cell's DNA (chromosomes) is doubled and partitioned equally between two daughter cells (Fig. 13.4). This was first observed about 200 years ago when microscope lenses had lower resolution, and so biologists could only observe mitosis, the period when the cell begins to stretch and break in the middle to form the two daughter cells. In mammalian cells, the time between mitotic events (M) is about 20 hours. Since nothing was visibly evident in this time interval, it was named Interphase (I).

As microscopes became better, biologists could also distinguish the S phase, when new chromosomes are being synthesized, as the chromosomes are sufficiently dense to be visible with better microscopes. Then the intervals between *M* and *S* were seen as Gaps. The gap phases are not silent, inactive steps in the cycle; rather, they have important functions in preparing cells for the *S* and *M* phases and, importantly, provide checkpoints to halt the cell cycle if something goes wrong during *S* or *M*.

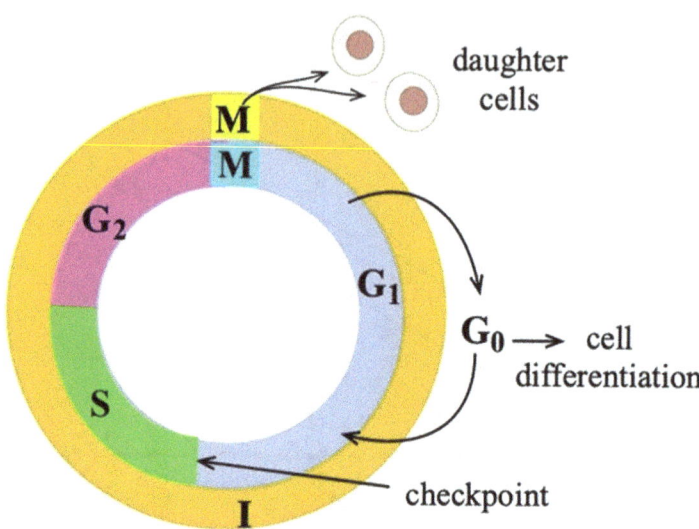

Fig. 13.4 The cell cycle. *G* Gap, *S* Synthesis of DNA, *M* mitosis, *I* Interphase

The Cell Cycle is an amazing process that drives a cell to make a replica of itself, often in less than an hour (for bacterial cells). For human cells, replication requires duplicating more than 3 billion base pairs of DNA, ideally without making a single mistake in a DNA coding region! For proper maintenance, the cell cycle needs controls, so that growth and division do not occur at the wrong time, are not too rapid nor too slow, and stop when required. Failure of these controls can cause disease, in particular, cancer. Cancer results from cells that should not divide but are caused to divide uncontrollably, producing tumors that invade healthy tissue and impair their functions. Thus, it is vital to understand how the cell cycle operates and is regulated.

Most cells in a fully developed animal stop proliferation and enter a nondividing state called G_0, where they may remain for weeks or even the lifetime of the organism. Such cells, however, are metabolically active. G_0 cells can re-enter the cell cycle, but this re-entry is regulated, providing control of cell proliferation. Cancer may also result from a G_0 cell that inappropriately re-enters the cell cycle.

How do cells "know" when to enter the various stages of the cell cycle, and what keeps the cycle moving forward? The entry of cells into the S or M phase is initiated by enzymes called *cyclin-dependent protein kinases* (Cdks). Cdks control the activities of proteins necessary to bring about the various events in the cell cycle (e.g. initiation of DNA synthesis, mitosis, cytokinesis). By themselves, however, Cdks are inactive; they must bind a partner protein called a *cyclin* (which is why they are cyclin-dependent) in order to become enzymatically active kinases. Cyclins are so named because their concentrations rise and fall during the cell cycle, and, in turn, so do the activities of the Cdks, thereby bringing about the various cell cycle transitions.

To understand how these regulatory proteins function, let us first consider the normal cell cycle. Nondividing cells are in the G_0 phase (Fig. 13.4), which is the phase in which cells normally differentiate—that is, they begin to express the genes that code for the proteins that define these cells for their special tissue properties, such as liver, or muscle, etc. Cells that remain mitotic are mostly in the G_1 phase. In this phase, they can accumulate slightly higher concentrations of nucleotides and amino acids in preparation for the S phase, where the chromosomes will be duplicated, as well as all the important histones. The latter act as a protein framework, forming something like a spool around which the DNA strands will be wrapped.

Uncontrolled cell division, leading to cancer, is most commonly produced when a mutation occurs to alter the function of two types of regulatory proteins: *oncogenes* and *tumor suppressors*.

2   Tumor Suppressors

In normal cells, tumor suppressors are responsible for:

- Suppression of cellular proliferation
- Induction/maintenance of differentiation
- Initiating cell death in response to stress or damage (Fig. 13.5).

> Examples: Rb gene: normally prevents retinoblastoma (eye cancer).
>
> p53 gene: a general tumor suppressor; most commonly mutated gene in human cancers.

They can trigger *apoptosis* (Fig. 13.5), which is a programmed cell death, before genetic abnormalities are inherited. Apoptosis should not be pronounced as: ay-POP-toesis; better is: ap-*uh*-**toh**-sis, from the Greek: "ptosis" meaning a drooping or falling off, defending against cancer by causing the cell to "fall off" or die. Readers are probably familiar with other words of Greek origin where the "*p*" is silent: pneumonia, psychosis, etc.

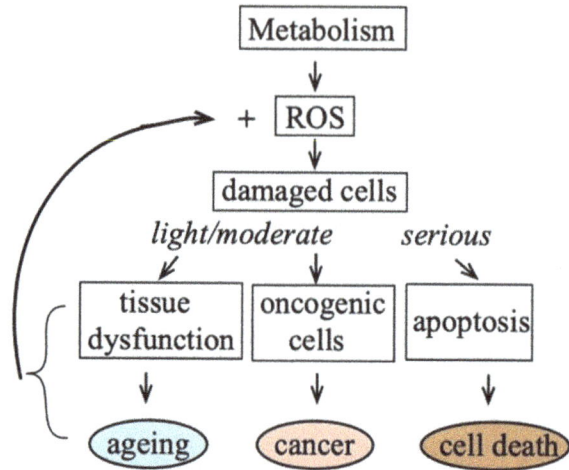

Fig. 13.5 Fate of damaged cells. *ROS* reactive oxygen species

3 A Mutated Rb Tumor Suppressor Leads to Retinoblastoma

An important regulatory factor for transcribing DNA is *E2F* (early region 2 binding factor), which helps to attract RNA polymerase to the 5′-end at the TATA box (Fig. 13.6). E2F actually binds at regulatory sites that are much farther 5′, but can still help the polymerase bind at the TATA box.

To prevent it from performing this function, a *tumor suppressor, Rb,* binds to E2F and blocks its function. Rb was so named because it was first discovered to prevent retinoblastoma in children. Most cancers are internal and more difficult to detect, but retinoblastoma frequently occurs in childhood. The expansion of the retinal tumor makes the eyes cloudy and leads to blindness, making detection easier.

Let us examine Fig. 13.7: Fig. 13.7a shows the normal function of E2F in attracting the polymerase, which then leads to the transcription of the adjacent gene—the ORF (open reading frame). The regulatory regions where E2F binds are connected to genes that code for proteins required in DNA synthesis. Figure 13.7b, at left, shows how the suppressor protein, Rb, normally binds with E2F, thereby preventing any transcription.

Binding of Rb to E2F maintains this gene in the off mode. This can be undone by a *cyclin-dependent kinase* (Cdk), which is able to phosphorylate the Rb protein, causing it to release its binding to E2F (Fig. 13.7b). Then

13 Traitors: Cancer Results When Regulatory Proteins Become Mutated

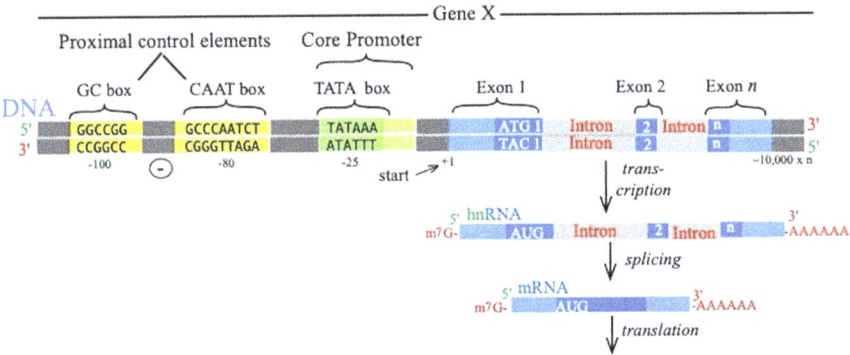

Fig. 13.6 A gene with the control regions for regulating transcription. The coding region is to the right; the regulatory factor binding sites are to the left (5′ end)

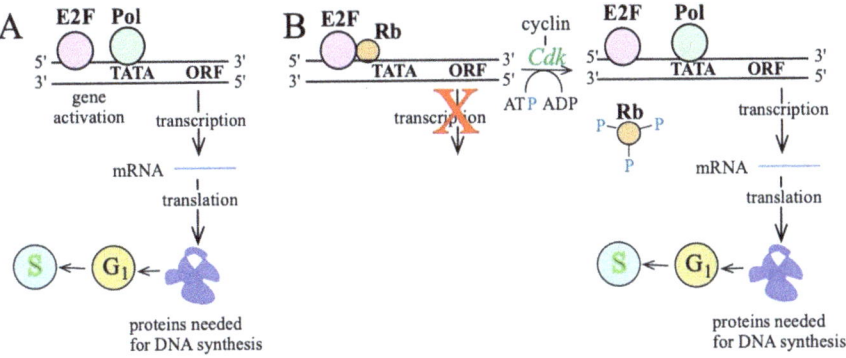

Fig. 13.7 The role of Rb in controlling genes required for cell cycle progression. (a) E2F initiates transcription by attracting a polymerase. (b) Unphosphorylated Rb is bound to E2F during most of G_1. The E2F-Rb complex binds to regulatory regions of numerous genes and prevents their transcription. Activation of *Cdk* causes phosphorylation of Rb and its dissociation from E2F. E2F is now able to stimulate transcription of target genes that encode proteins necessary for cell cycle progression. *ORF* open reading frame with the codons for the protein to be synthesized

transcription and translation proceed unchecked. But, Cdk only functions when it is activated by cyclins, which are often produced in cells that are in the G_1 phase. This disease requires only two mutant copies of the suppressor protein. If a child of a heterozygous parent, who has already inherited one mutant gene, then has a new mutation in the good gene, retinoblastoma will soon occur. This is an example of "one hit."

4 p53 Tumor Suppressor

While Rb was the first tumor suppressor to be identified for this function, p53 may be more important, as it appears to be the most widely used protein with this function (Fig. 13.8). How was p53 named? We might remember from Chap. 1 that some proteins were first identified by being detected with electrophoresis. For this figure, small tumor samples were obtained from patients; proteins were extracted and analyzed on *western blots* (see Fig. 13.9 for an example) with antibodies specific to the p53 protein. Such antibodies had been prepared earlier and were now used to probe for the presence of complete p53 protein molecules (Fig. 13.9).

Three of the tumor samples had no p53 protein. Patient 14 had some p53 at the expected size but also had a protein at a much larger size recognized by the antibody. This larger version of p53 could result from:

1. A new splice site in an intron, by which part of the intron is now included in the final mRNA, thereby making it code for a larger protein.
2. A mutation in the termination signal, so that transcription would go beyond the end of the normal gene, producing a larger mRNA and then a larger protein.

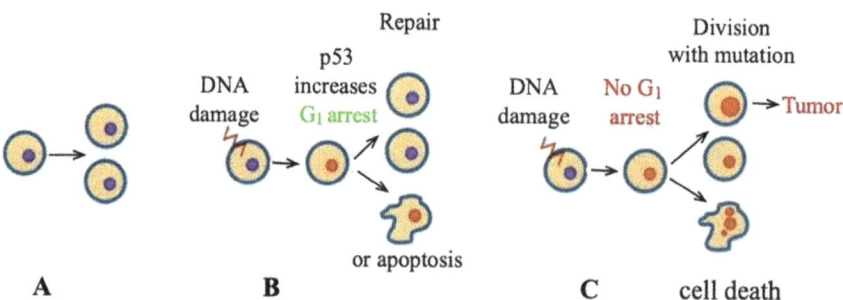

Fig. 13.8 Functions of the p53 tumor suppressor. (a) Normal cell division. (b) DNA is mutated and p53 concentrations increase in response, and result in repair or apoptosis. (c) DNA is mutated after p53 has been mutated, therefore no repair or apoptosis, and a tumor results

13 Traitors: Cancer Results When Regulatory Proteins Become Mutated

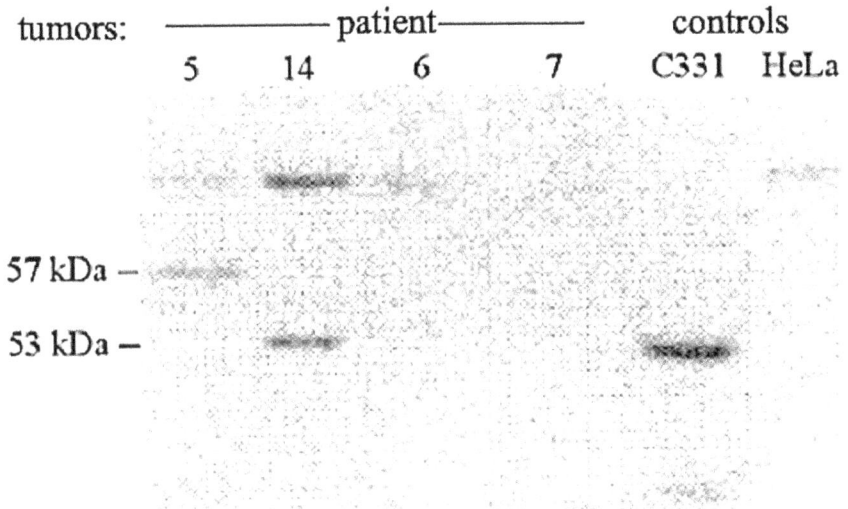

Fig. 13.9 A western blot (Also Called Immunoblot) to Test for p53 Protein Expression. The first four lanes have patient samples. C331 and HeLa are human transformed cell lines that are used as controls: C331 always expresses p53, and HeLa normally does not. p53 is expected to be located at 53 kDa. (Source: Proc. Natl. Acad. Sci. USA (1990), 87: 5863–5867)

Why did scientists describe the experiment in Fig. 13.9 as a *western blot*? The first effort to detect DNA bands on an electrophoresis blot was performed by an English scientist, Edwin Southern, in 1975. He isolated DNA fragments from sample tissues, separated them by electrophoresis, then transferred them onto a filter paper, and probed them with radioactive DNA fragments that were complementary (i.e. they would bind with) the expected target DNA he was looking for. After he published his results, many scientists used his procedure and simply referred to it as "Southern blotting" to give credit to the person who originated this procedure. Then, other scientists soon employed this procedure but used complementary RNA fragments as probes instead of complementary DNA, and, with a sense of humor, distinguished this slightly altered process as "Northern blots." By this time, the ability to raise antibodies against clinically important proteins also made it possible to separate proteins by electrophoresis and then use specific antibodies to probe for them, and these were then called "western blots." Scientists also have a sense of humor.

5 Oncogenes

Oncogenes (from Greek *onco* = tumor, plus gene) are any genes, or their proteins, that have the potential to promote tumor formation. The normal version of such proteins is called a *proto-oncogene*. The three Ras genes in humans (HRAS, KRAS, and NRAS) are the most common oncogenes in human cancer. The *ras* oncogene encodes a constitutively active form of the Ras protein that cannot shut off because it has lost the ability to hydrolyze GTP efficiently. Ras functions as a G-protein, as described in Chap. 8, and see Fig. 13.10.

Because the mutant ras protein activates cell division without itself being activated by an external signal, it has a dominant effect, meaning that only one of the two ras genes needs to be mutated for cell division to proceed without stop.

An overall sequence of mutation events, involving ras, and leading to colon cancer, is shown in Fig. 13.11. It actually has six steps, or "hits," among which is the other tumor suppressor—p53. It has not been verified that all six hits in the sequence of steps are essential to produce cancer. Since both tumor suppressors have been mutated, one or two of the other hits may not have been necessary. The sequence may represent a slightly fortuitous time sequence, with the mutated p53 being the final blow.

Oncogenes stimulate cell division and therefore are dominant, meaning that only one gene needs to be mutated to produce uncontrolled cell division

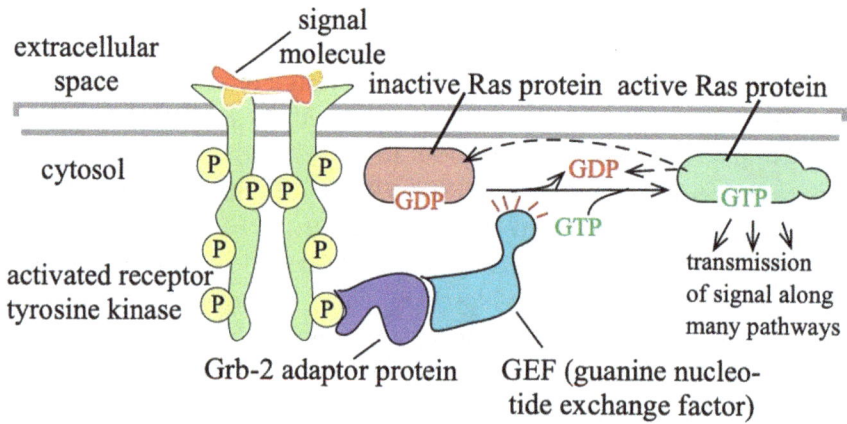

Fig. 13.10 The active conformation of Ras (green) requires binding of GTP. After slow hydrolysis of GTP to form GDP, the enzyme changes to very low activity. A guanine nucleotide exchange factor (GEF) promotes release of tightly bound GDP, so that GTP can again bind and activate the enzyme. GTPase-activating protein (GAP) replaces GEF and speeds up hydrolysis of GTP and inactivation of the enzyme

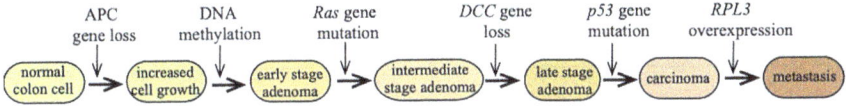

Fig. 13.11 The sequential accumulation of mutations in tumor suppressor and oncogenes giving rise to metastatic colon cancer

leading to cancer. In contrast, p53, as a suppressor, must have both genes mutated so that both copies are nonfunctional to make uncontrolled cell division possible.

6   Viruses Can Cause Cancer

Some viruses can also be involved in promoting cancer. This happens because their genome, if DNA, can be inserted directly into the host cell's DNA, and if RNA, has to be reverse transcribed into a DNA copy, which can then be inserted into the host cell's DNA (see Fig. 12.1). Where these viral genes are inserted is a matter of chance, and sometimes they will be adjacent to some of the human cell's genes that code for proteins in cell division. Because viral genes always carry a strong promoter, which attracts TFs to assure continuous expression of the viral genes, so as to make many viral proteins, if they are inserted near host cell genes that are used for cell division, then these genes will also be activated for continued expression. Then host cells will become mitotic without control.

Quite a few viruses have by now been shown to be involved in the formation of cancers in humans. Table 13.1 lists some of the more important examples. It should be noted that having one of these viruses does not always lead to cancer because of the randomness of where the DNA is inserted in human cells.

Good examples of this are the papova viruses, which have been demonstrated to cause oral and genital cancers. The human papilloma virus is a double-stranded DNA virus that infects the epithelial cells of skin and mucosa. Of the more than 120 strains of HPV, about a dozen cause warts or cancers in skin/epithelial tissue, and new studies show that HPV16 and HPV18 are also linked to oral and genital cancer.

Oral cancers comprise 2–3% of all cancers in men and women, and are spread throughout the oropharyngeal region. Dentists and oral hygienists are taught to search for and recognize this cancer at the back of your throat. About 90% of oral cancers are squamous cell carcinomas and result in *an*

Table 13.1 Viruses associated with human cancers

Virus	Associated tumors
DNA viruses	
Papova virus family	
Papilloma virus or HPV (many distinct strains)	Pharyngeal cancer; cervical cancer; warts (benign)
Hepadnavirus family	
Hepatitis B virus	Liver cancer (hepatocellular carcinoma)
Herpesvirus family	
Epstein-Barr virus	Burkitt's lymphoma
RNA viruses	
Retrovirus family	
Human T-cell leukemia Virus type I (HTLV-I)	Adult T-cell leukemia/lymphoma
Human immunodeficiency virus (HIV-1)	Kaposi's sarcoma; White blood cells (lymphocytes)

overall mortality rate of about 50%. Surveys in 2010 suggested that about 50% of college students tested positive for this viral DNA. Because the actual rate of cancer development is so low, the behavior of students in terms of sexual experimentation or hygiene appears to be unchanged, and so the virus continues to be widespread in the younger generations.

The hepadna viruses that can cause hepatitis B are also potentially dangerous.

The *retroviruses*, so named because they have an RNA genome that must be reverse transcribed into DNA, are associated with T-cell leukemia and lymphoma. The importance of HIV in destroying T-cells was described earlier in this chapter. However, HIV does not cause cancer itself. It is correlated with higher cancer rates in victims because infected people have a compromised immune system and therefore cannot as easily detect and remove cancer cells before they metastasize.

Resources

Trisomy 21: https://www.chop.edu/conditions-diseases/trisomy-21-down-syndrome

Tumor growth: https://www.cancerresearchuk.org/about-cancer/what-is-cancer/how-cancers-grow

Cell cycle: https://en.wikipedia.org/wiki/Cell_cycle

Reactive oxygen species: https://en.wikipedia.org/wiki/Reactive_oxygen_species

Rb tumor suppressor: https://pmc.ncbi.nlm.nih.gov/articles/PMC5228373/
p53 tumor suppressor: https://en.wikipedia.org/wiki/P53
3 types of blots: https://www.labmanager.com/southern-vs-northern-vs-western-blotting-techniques-854
Ras GTPase: https://en.wikipedia.org/wiki/Ras_GTPase

Epilogue

We have now had a reasonable survey of a variety of proteins, enzymes, and their cousins—the regulatory proteins, transcription factors, tumor suppressors, hemoglobin, special viral reverse transcriptase, etc. While we still remain unsure as to the total number of enzymes in humans, we do know that there are about 5000 different metabolic reactions, which are almost all performed by enzymes, and perhaps 1000 functions performed by the other types of proteins mentioned. Because of the need for isozymes, which perform the same chemical reaction, but have been slightly mutated to work a little bit better in some different tissues, the total number of enzymes in a human is close to 20,000.

We now know that a special feature of all proteins is that they have the ability to bind to one or more other proteins, usually their own type, to make oligomers (from Greek: many bodies = a complex containing two or more of the same type of enzyme subunit), and sometimes to regulatory subunits. In addition, all of these bind to some normal metabolite or cellular compound (e.g., DNA). Upon binding, they either chemically alter this substrate (i.e., enzymes) or produce a conformational change in the bound ligand (e.g., activators of cell cycle enzymes).

We have seen the importance of conformational change when it is accompanied by a change in affinity for the bound ligand. That is why the conformational change in our hemoglobin has evolved to be appropriate for first binding oxygen in the lungs with higher affinity for oxygen and then weaker when releasing it in the capillaries of the tissues that need this oxygen.

We have seen how a sequence of chemical modifications, usually phosphorylation, can activate an enzyme cascade. Examples of this are in the hormone-activated release of glucose from stored glycogen and also in the blood clotting cascade, where a series of specific proteases activate their target enzyme substrates, so that a blood clot can form within a few minutes or less.

We also have seen the benefit and purpose of duplicated genes that produce isozymes. Among these are the many collagens, the different globins, the various hexokinases, etc. With the hexokinases, we learned that the four isozymes perform the same chemical reaction, but by being different isozymes with altered affinities for glucose and by being uniquely expressed in those tissues where this altered affinity is beneficial, the presence of a unique isozyme becomes important for the tissues. Therefore, Hexokinase I and Hexokinase III, with the best affinities for oxygen, are in our RBCs. RBCs have no mitochondria and are, therefore, totally dependent on glycolysis for energy. With these isozymes, our red blood cells can still use the available glucose, even if a person becomes hypoglycemic.

Of course, the first thing that we learned is how fast enzymes can be. The slowest metabolic enzymes still have a k_{cat} of 1 s^{-1}. Most metabolic enzymes have rates between 100 and 500 s^{-1}. And the Olympic champion for speed is carbonic anhydrase, with a rate of about 1 million reactions per second.

So, like the thousands of worker bees maintaining their beehive, the almost 20,000 different enzymes in humans make it possible not just to have a collection of living cells but also, by the occurrence of isozymes and other isoproteins, to have many specialized tissues, which together make our much more complex bodies function so well.

We should then be all able to appreciate a simple fact: enzymes, and their kindred proteins, are amazing! Without them, life would not be possible.

Index

A

Abiotic, 44–46, 48, 55–57
Accelerate, 13
Acceptor, 12
Acetate, 50, 56
Acetoacetate, 177
Acetyl, 71, 73, 178
Acetylation, 73
Acetyl–CoA, 173, 177, 179
Acid, 3, 12, 15, 16, 22, 23, 29, 31–33, 44, 45, 50, 55–57, 59, 63–67, 69, 71–77, 81, 95, 100, 105–135, 144, 151, 160, 161, 165, 169–171, 175–179, 183, 185, 187, 189, 203
Acidic, 6, 20, 81, 90, 130, 175
Acidity, 6, 26, 27, 88, 90, 182
Acrobats, 69, 70
Actin, 14, 15, 62
Activated, 28, 96, 98, 100, 101, 148, 166, 170, 174, 205, 208, 209
Activator, 69, 96, 98, 103, 172, 173
Active site, 3, 10, 71, 73
Activity, 3, 5–7, 12, 18, 28, 30, 73, 98, 101, 102, 104, 113, 115, 129, 130, 132, 161, 162, 165, 170–172, 174, 178, 202, 203, 208
Adenosine triphosphate (ATP), 11, 12, 15, 20, 29, 41, 42, 48, 50, 52, 58, 82, 85, 91, 95, 110, 118, 163, 164, 167–175, 179, 180, 185
Affinity, 7–10, 29, 83, 84, 87–91, 94–98, 146, 165–167, 170, 173, 174
Africa, 45, 46, 142, 143, 147, 148, 189
Agent, 4, 5, 11, 20, 26, 27, 82, 139–141, 182, 184, 187, 189
Air, 9, 39, 42, 43, 80, 83–85, 183, 193
Airplane, 14
Alanine, 56
Albumin, 15, 22
Alcohol, 4, 11
Algae, 2, 42
Align, 28, 112, 124, 153, 172
Alignment, 3, 64–66, 138, 140, 147, 170
Allosteric, 93–104, 162, 168
Alphabet, 45
Alpha-carbon, 44, 67
α-helical, 62

Index

Alpha helices, 63
Altitude, 53, 83
Ambivalent, 66
American, 17, 22, 30, 107, 147, 160, 189, 194
Amino acids, 3, 22, 29, 31, 32, 44, 45, 50, 55–57, 59, 63–67, 69, 71–77, 95, 100, 116, 118, 124, 127–130, 144, 151, 160, 165, 170, 177, 185, 187, 189, 191, 203
  alpha, 44
  beta-, 44, 45
  essential, 56
  nonessential, 56
Amino group, 56, 75
Ammonium, 56
Amoebas, 2
Anabolism, 50
Anaerobic, 42, 174
Anatomy, 80, 151
Ancestor, 57, 58, 106, 111, 142, 143, 170
Anemia, 41, 128, 151–157, 198
Animals, 1, 20, 43, 49, 102, 118, 134, 138, 143, 147, 161, 182, 202
Anoxic, 83, 152
Antibody, 75, 139–142, 183, 187, 188, 193, 206, 207
Anticodon, 76, 77
Antigen, 15, 140, 141, 201
Antiparallel, 107, 108, 115
Aorta, 80
Apple, 9–11, 85
Aqueous, 45, 54, 66
Arabidopsis thaliana, 111
Archaea, 57
Archaebacteria, 42, 43
Archeological, 65, 147
Arginine, 31, 32, 56
Arginine decarboxylase, 32
Arrow, 33, 52, 54, 62–64, 72, 96, 145

Arterial, 157
Arteries, 62, 80, 102, 154, 192
Ascorbate, 45, 56
Asparagine, 56
Aspartate/aspartic acid, 56
Assembly, 2, 3, 9, 38, 45, 48, 55, 95
Assembly line, 2, 3, 9
Astronomical, 31, 38
Atmosphere, 37, 42, 43, 53, 84, 147
Atom, 12, 26, 33, 38, 40, 41, 44, 45, 49, 55, 59, 63, 69, 75, 77, 80, 108, 148, 149, 163, 169, 179
Atomic, 7, 12, 40, 41, 63, 64, 108
AUG, 75–77, 117, 119–121
Auto, 11, 14, 142
Autoimmune, 142
Automated, 2, 121
Automatically, 10, 11, 15, 132
Automobile, 2
Autopilot, 10, 11
Autosomal, 127, 138
Avogadro, 7, 27

B

Backbone, 44, 45, 55, 63, 67, 74, 108, 109, 131, 179
Bacteria, 2, 14, 20, 28, 42, 43, 45, 57, 76, 82, 90, 102, 106, 109–111, 113, 166, 182, 184, 185
Bacterial, 2, 15, 17–20, 28, 45, 56–58, 106, 113, 114, 184, 185, 202
Bacterium, 14, 32, 114, 139, 185
Bandage, 102
Barometer, 84
Base, 15, 22, 23, 33, 35, 44, 77, 107–109, 112, 116, 124–131, 134, 138, 202
Baseball, 11
Based, 61, 66, 126, 170, 193
Base pair, 35, 77, 107–109, 112, 116, 130, 131, 202

Battle, 102
Battlefields, 103
Beer, 11
β-pleated sheets, 62
Beta strands, 63
Big Bang, 40, 45
Binding, 3, 10, 15, 27, 29, 49, 69, 77, 80–82, 84–91, 94–96, 103, 104, 119, 132, 135, 140, 141, 152, 153, 156, 162–165, 172, 173, 192, 204, 205, 208
Biochemical, 14, 18, 30–32, 118, 185, 200, 201
Biochemistry, 33–35
Biosynthesis, 11, 91
Bisphosphoglycerate (BPG), 88, 90
Blood, 6, 15, 16, 20, 21, 29, 30, 41, 49, 73, 79–91, 96–102, 141, 151–154, 156, 160–163, 165–167, 171, 177, 183, 188, 189, 192, 200, 210
Blood clot, 97, 99
Blood clotting, 21, 41, 97–102
Body, 2, 6, 7, 17–23, 29, 39, 41, 62, 69, 70, 80, 81, 85, 89–91, 130, 133, 140, 142, 143, 150, 163, 165, 167, 176, 177, 179, 182, 192, 200, 206
Body fluid, 6, 29
Bond, 12, 22, 26, 29, 32, 38, 44, 55, 56, 59, 63, 64, 66, 67, 74–75, 77, 100, 107, 108, 112, 131, 161, 163–165, 179
Bone, 16, 57, 62, 147
Book, 6, 15, 80, 108
Brain, 28, 81, 83, 98, 116, 130, 134, 160, 162, 165, 167, 169, 171, 177, 193
BRENDA, 15
Bronze Age, 102
Buffer, 6–7, 122, 152, 153
Building blocks, 4, 44–46, 49
Bya, 40
By-product, 42, 90, 174

Caenorhabditis elegans, 111
Calcium, 6, 41, 101, 175, 176
Cambridge, 108
Cancer, 13, 15, 43, 107, 130, 132, 143, 146, 197–210
Cannery, 9
Capillary, 21, 49, 80, 81, 85, 87, 88, 90, 91, 94, 96, 102, 152, 154, 156, 192
Capsid, 35
Carbon, 12, 33, 41, 44, 45, 55, 58, 59, 67, 90, 144, 145, 148, 161, 163, 167, 169, 170, 178, 179
Carbon dioxide (CO_2), 7, 43, 48, 80, 81, 85, 90
Carbonic anhydrase, 16, 27, 31, 32
Carboxyl, 71, 75, 175, 177
Carrier, 15, 16, 33, 85
Cartilage, 16, 62
Cascade, 21, 41, 97, 99, 100, 159, 164, 165
Catabolism, 50, 51
Catalase, 27
Catalyst, 4, 5, 13, 30
Catalytic, 3–5, 10, 13, 14, 26–28, 71, 98, 101, 104, 162, 163, 170, 173, 191
Catalyze, 32, 77, 100, 112, 132, 163
Cats, 45
Cell, 2, 4–6, 10, 11, 13–23, 25–30, 33, 35, 37, 41–45, 47–59, 67, 72, 73, 80–83, 85–91, 95, 96, 100, 101, 104, 106, 109, 111, 113, 114, 116, 118, 122, 127–130, 134, 140, 142–144, 146–149, 151–157, 161–165, 167, 170, 171, 173–175, 182–189, 191–193, 198–210
Cell death, 28, 203
Cellular, 2, 4, 14, 20, 27, 42, 43, 48, 50, 52, 61, 72–75, 85, 91, 104, 163, 174, 175, 179, 184, 185, 199–201, 203, 210

Central dogma, 111
Century, 2, 4, 6, 17, 46, 151, 181, 189
Charge, 73, 75, 122, 153
Chemical, 2–7, 10, 12–15, 18, 25–28, 30–32, 40, 42, 44, 45, 47, 48, 50–56, 59, 67, 71–73, 121, 122, 127, 130, 143, 160, 175, 185, 188, 198, 200, 201
Chemically, 3, 4, 15, 26, 42, 44, 53, 98, 116, 160
Chemist, 44, 52, 55, 58
Chemistry, 5, 33–35, 37, 58, 62–64, 71, 82, 143, 149
Chick, 22
Chimpanzee, 133, 134
Chlorophyll, 42, 48–50
Chromosome, 20, 106, 109, 114, 116, 121, 126, 138–142, 191, 198, 201, 203
Circulatory, 79–91, 101
Citric acid, 169, 170, 175, 176, 179
Citrus, 45
Clot, 97–99, 101
Coast, 39, 147
Cobalamin, 49
Cobalt, 41, 49
Code, 2, 13, 14, 16, 22, 45, 56, 57, 65, 66, 69, 70, 72, 76, 77, 109, 116, 118, 130, 140, 170, 191, 203–206, 208, 209
Codon, 64, 75–77, 116, 117, 119, 120, 124, 127–130, 170, 205
Collagen, 14–16, 62
Comet, 38
Companies, 9, 96, 124
Competency, 186
Competitive, 3, 11
Complementary, 107, 116, 120, 138, 207
Complementary strand, 109, 116, 131, 138
Complex, 1–3, 5, 6, 19–21, 38, 43, 44, 48, 49, 53, 68, 73, 96, 106, 107, 110, 116, 118, 142, 161, 205

Compound, 3, 4, 6, 7, 26, 29, 30, 32, 44, 48–55, 57–59, 73, 108, 144–146, 150, 157, 161, 170, 172, 173, 175
Concentration, 5–8, 10, 29, 30, 42, 43, 51, 52, 54, 81, 83–87, 94, 95, 98, 100, 101, 104, 145, 152–154, 157, 160, 161, 163, 165, 167, 170–174, 203, 206
Conform, 47
Conformation, 69–71, 74–75, 94–96, 98, 103, 104, 152, 164, 172, 208
Constituent, 18–20, 48
Conveyor, 9–11
Cooperativity, 95, 168–171
Cornea, 62
Covalent, 73, 75, 100, 161, 162, 164
Covid, 22, 35, 181, 191–194
Crick, F., 107, 108
Cross-linked, 100
Cryptic, 13
Crystalline, 32
Crystallize, 53
Crystal structure, 107
Curve, 8, 86–91, 94, 95, 133, 171, 172
Customer, 51, 52
Cyanobacteria, 42, 43, 48
Cycling, 25
Cygb, 81, 83, 85
Cysteine, 56
Cytoglobin, 81, 83, 85
Cytoplasm, 18, 19, 29, 71, 73, 75, 99, 103, 162
Cytoplasmic, 73

D

Daltons, 69
Damage, 20, 25–28, 35, 43, 73, 98, 106, 126, 130, 131, 161, 182, 203, 204
Daughter, 201

Deficiency, 41, 149
Dehydrogenase, 16, 69, 168, 170, 174–176
Dehydrogenate, 12
Deleterious, 46, 98, 126, 151–155
Deliver, 94
Delta G, 52, 53
ΔK_M, 98
Deoxygenated, 151, 152, 154, 155
Deoxy-hemoglobin, 155
Deoxynucleotide, 110, 113, 115, 116
Deoxyribonucleic acid (DNA), 2, 13, 15–18, 20, 22, 26–28, 32, 33, 35, 43, 57, 64, 67–70, 75, 106–135, 138–140, 142, 147, 149, 181, 184, 185, 187, 191, 193, 201–204, 206–207, 209, 210
Deoxyribose, 33
Development, 14, 127, 134, 142, 154, 210
Diabetes, 142
Diagram, 10, 19, 30, 32, 45, 50, 54, 55, 63, 64, 72, 75, 116, 119, 168, 169, 191
Diffuse, 21, 42, 49, 80, 85, 118
Digested, 166, 167
Digestive, 5, 6, 100, 130
Dimension, 23
Dimer, 28, 68, 69, 84
Disorder, 13, 101, 132, 188, 198
Dissociate, 26, 77, 91, 95
Disulfide, 74, 75
Doctor, 149, 150
Doctoral, 107
Dogs, 45
Domain, 64, 65, 69–71, 73, 130, 146, 173, 192
Donor, 12, 101, 189
Double helical, 35
Double helix, 22, 107, 110
Downhill, 52, 53
Drosophila melanogaster, 111

Duplex, 138, 139
Duplicated gene, 15, 16, 134, 170
Dynamic, 48, 51

E

Earth, 30, 32, 33, 37–46, 48, 57
EcoR1, 28
Effect, 26, 51, 95, 97, 127, 130, 155, 208
Effector, 86, 88–91, 94, 97, 101, 102, 104
Egg, 9, 21, 22, 106
Electricity, 52, 122
Electrostatic, 75
Element, 11, 12, 37, 39–41, 45, 55, 58, 62, 118
Elephant bird, 22
Embryo, 14
Employees, 96
Emu, 22
Encode, 205, 208
Endoplasmic, 72
Energy, 1, 2, 12, 15, 20, 26, 27, 29, 32, 37, 42, 48–50, 52–55, 75, 81, 89, 96, 118, 139, 149, 160, 161, 163–165, 167, 171, 173, 175–177, 179
Energy barrier, 47, 53, 54
England, 108
Englishman, 107
Environment, 14, 18, 37, 48, 66, 126, 147, 150, 200
Equilibria, 48, 51–58, 95
Error, 27, 45, 126, 127, 129–131, 142, 187
Erythrocyte, 21, 81–83, 91
Escherichia coli, 14, 18, 19, 67, 106, 109, 111, 113, 187
Ethane, 50, 75
Eukaryote, 42, 43, 57, 58, 110, 118
Eukaryotic, 19, 20, 58, 67
Evaporate, 53, 143

Evolution, 3, 10, 15, 25, 29, 33–35, 41, 45, 48, 50, 134, 137–157, 160, 163, 164, 170, 172
Evolutionary, 57, 58, 67, 71, 109, 138, 151
Evolved, 4, 16, 20, 42, 45, 58, 81, 82, 85, 88, 91, 95, 98, 100, 107, 109, 132, 165, 178, 187, 192
Exodus, 147
Exon, 69, 70, 116–118, 145, 146
Exonuclease, 113, 115, 118, 129, 130
Exponential, 26, 27, 185, 186
Exterior, 72
External, 2, 45, 50, 52, 66, 85, 103, 208
Extracellular, 72–75, 104, 179
Extrusion, 72
Eye, 17, 62, 144, 146, 182, 203, 204

F

Factor, 4, 15, 74–77, 100, 101, 103, 104, 149, 157, 204, 205, 208
Factor VIIa, 100
Factor IX, 100
Factor X, 100, 101
Factor XI, 100
Factor XII, 100
Factory, 2
Fat, 29, 50, 90, 91, 98, 134, 160, 163, 165, 167, 171, 177–179
Fathers, 198, 201
Fatigue, 161
Fatty acid, 15, 73, 161, 177–179
Fellowships, 108
Fermentation, 4, 11
Fermented, 4
Fetal, 16, 83, 89–91, 127
Fetus, 16, 83, 90, 91
Fever, 90
Fibrin, 97, 100–102
Fibrinogen, 62, 100, 101
Fibrous, 61

Filaments, 18, 62
Flagellum, 18
Folding, 64–66, 69, 71, 128
Frameshift, 127
Franklin, R., 107
Free energy, 52, 53, 96
French, 4–6
Frequency, 43, 119, 124, 142, 149, 150
Fruit fly, 111
Function, 1–3, 5, 10, 13, 15–18, 20–23, 27, 30, 37, 41, 43, 47–59, 64, 66, 68, 71–73, 80, 81, 87, 95, 98, 100, 101, 110, 111, 118–120, 127, 129, 132, 134, 138, 142–144, 146, 148, 151, 153, 160–165, 167, 170, 171, 173, 184, 185, 188, 191, 200–206, 208
Fusion, 40, 101, 183, 189

G

ΔG, 52–54
$\Delta G\ddagger$, 53
Gap, 113, 131, 201, 202
Gas, 37, 38, 40, 43, 54, 83, 84, 152, 166
Gb, 22, 23, 94, 95
GDP, 74, 103, 104, 208
Gene, 2, 13, 15, 45, 57, 58, 69, 91, 98, 106, 109, 116–120, 128, 134, 135, 141, 142, 145, 146, 151, 154–157, 170, 184, 189, 191, 198, 203–206, 208
Gene duplication, 15, 57, 91, 170
Genetic, 22, 33, 45, 57, 109, 111, 126, 127, 148, 182, 184, 198, 203
Genome, 22, 23, 28, 35, 66, 109, 113, 114, 121, 122, 124, 129, 138, 139, 185–187, 189, 191, 209, 210
Geologists, 26
German, 5–7, 147

Gibbs, W., 52
Giga, 23
Globin, 82–87, 90, 91, 94, 128, 151
Globular, 61
Glucokinase, 12
Glucose, 4, 5, 12, 29, 30, 50, 69, 73, 81, 159–180
Glutamate, 56, 128, 153, 165
Glutamic acid, 151
Glutamine, 56, 100
Glyceraldehyde, 12, 13
Glycine, 55, 56
Glycogen, 28, 69, 161–165, 167, 173, 177
Glycogen phosphorylase, 28, 69, 161, 163, 164
Glycolipid, 185
Glycolysis, 157, 167–176, 179, 180
Glycoprotein, 13, 185
Goldilocks Zone, 38, 39, 43, 172
gp42, 13
G-protein, 73, 74, 98, 103–104, 208
G-protein-coupled receptor (GPCR), 73, 103, 104
Grams, 7, 30, 133
Gravitational, 39, 40, 53
Gravity, 38
Greek, 4, 5, 12, 15, 19–21, 42, 45, 51, 57, 82, 95, 106, 118, 122, 142, 150, 151, 163, 184, 198, 203, 208
Growth, 15, 22, 133, 185, 198–200, 202
GTPase-activating protein (GAP), 104, 208
Guanine nucleotide exchange factor (GEF), 74, 103, 104, 208
Guanosine triphosphate (GTP), 13, 74, 93, 103, 104, 179, 208
Gut, 14, 18

H
HbA, 80, 81, 83, 85–91, 94–97, 152, 153, 156
HbF, 83, 89, 91, 155, 157
H-bond, 75, 108
HbS, 151, 152, 155–157
Heart, 80, 98, 161, 167
Heat, 30, 40, 51–53, 150, 182
Hebrew, 46, 143
Helicase, 112
Helices, 62, 63
Helium, 40
Helix, 22, 62, 63, 107, 110
Hematocrit, 82, 91
Heme, 49, 50, 84
Hemoglobin, 15, 16, 21, 41, 49, 69, 79–91, 94–97, 151–155, 157
Hemophilia, 101
Henri, V., 5, 6
Hexokinase, 16, 62, 69, 167, 169–172
Hg, 83
Histidine, 56
Histone, 15, 66, 135, 203
Homo sapiens, 66, 111, 142
Hormone, 15, 28, 74, 98, 103, 104, 134, 149, 163, 164, 192, 199
Host, 1, 28, 33, 182, 184, 186, 188, 191, 192, 209
Human, 2, 3, 6, 14–18, 20–23, 25, 41, 43, 45, 56, 67, 68, 82, 85, 101, 102, 110, 113, 120–122, 124, 129, 130, 133, 134, 138, 139, 142–148, 160, 167, 169, 171, 182–184, 187–189, 191–193, 202, 203, 207–210
Human immuno-deficiency virus (HIV), 22, 181, 183, 186–191, 210
Hydrogen, 6, 12, 22, 38, 40, 41, 63, 64, 69, 75, 108

Hydrogen bond, 22, 38, 63, 64, 75, 108
Hydrogen peroxide, 27
Hydrolase, 4, 12, 13, 100, 143, 163, 164
Hydrolysis, 51, 73, 103, 104, 208
Hydrolytic, 13, 20, 30
Hydrophilic, 66
Hydrophobic, 66, 71, 75, 128, 130, 153, 155

Ice, 11, 38, 39, 53, 54, 147, 148
Iceland, 197, 201
Immune, 139, 142, 161, 181, 184, 186–189, 191, 210
Immune system, 82, 139, 142, 161, 184, 187–189, 210
Immunoglobulin, 15, 141
Inactive, 71, 74, 96, 98, 100, 101, 103, 104, 162, 201, 202
Inanimate, 47, 48
Inch, 18, 122
Infect, 1, 182, 184, 186, 188, 189
Infection, 28, 35, 90, 109, 161, 182, 183, 186–188, 191, 192
Influenza, 22, 183, 187–188
Information, 13, 33, 57, 71–74, 116, 149, 184
Infusion, 101
Inhibitor, 3, 10–11, 69, 96, 164, 172, 173, 183
Injury, 20, 21, 82, 97, 99
Insulin, 69, 165, 166
Interior, 4, 20, 21, 66
Internet, 15
Intertidal, 39
Intestine, 81, 165, 166, 200
Intron, 105, 106, 109, 116–120, 127, 130, 146, 206
Ionic, 75
Iron, 41, 49
Isoleucine, 56

Isomerase, 55
Isoprotein, 15, 16
Isozyme, 14, 103, 134, 137, 143–146, 159, 165, 167–172

Jacob, F., 96
Joints, 62, 142
Journal, 5, 108, 147, 148
Junk DNA, 110, 121

k_{cat}, 25–27, 30–32, 98, 100–102, 134, 169
K_M, 7–9, 25, 29, 83, 84, 97, 101, 102, 134, 165, 167
k_{non}, 30–32
Kilobase, 22
Kilodaltons, 13, 69
Kilogram, 22, 133, 179
Kinase, 12, 41, 62, 74, 160–165, 172, 173, 197, 202–204
Kinetically, 12
Kinetics, 26, 169, 186
Kingdom, 57
King Solomon, 46
K-type, 93, 97, 98, 102

Lactate, 16, 69, 159, 168, 171, 174–176
Lactate dehydrogenase, 16, 69, 159, 168, 170, 174–176
Lactic acid, 81, 171, 175
Last Archaeal Common Ancestor (LACA), 58
Last Bacterial Common Ancestor (LBCA), 58
Last Eukaryote Common Ancestor (LECA), 58

Last Universal Common Ancestor (LUCA), 57, 58, 111
Latin, 15, 18, 20, 84, 116, 142, 163, 191, 199
Leader sequence, 72
Lethal, 46, 56, 90, 98, 157
Letter, 7, 13, 31, 52, 65, 76, 95, 129
Leucine, 56
Ligament, 62
Ligand, 15, 82, 84, 85, 94–96, 159, 162, 172
Ligare, 84, 116
Ligase, 116, 131, 137–139
Light, 4, 25, 27, 43, 54, 84, 117, 140, 141, 143, 144, 148–150, 171, 185
Limit, 25, 29, 47
Limited, 2, 21, 45, 55, 56, 93, 101, 103, 116, 118, 121, 124, 125, 137, 144, 165, 167, 177, 194
Lipid, 50, 59, 185
Liquid, 18, 38, 53, 83
Liver, 28, 134, 159, 161–163, 167, 171, 173, 177, 198, 200, 203
Loading, 86–88, 94, 95
Lungs, 21, 49, 80, 81, 83, 85–87, 94–96, 154, 188, 193
Lymphatic fluid, 41
Lymphocytes, 81, 181, 184, 200, 210
Lysine, 56, 100
Lysis, 12
Lysosome, 20

M

Macromolecule, 29
Macrophage, 82, 184
Magnesium, 6, 41, 49
Malfunction, 127
Malignant, 199
Mammalian, 6, 14, 19, 20, 35, 43, 105, 113, 166, 201
Mars, 38, 39

Maternal, 90, 91, 200
Matter, 10, 47, 48, 209
Mb, 81, 83, 85–88, 95
Meal, 6, 29, 73, 100, 162, 163, 166, 167, 176
Mechanical, 48, 51, 52
Medical, 87, 149–150, 191, 194
Megabase, 23
Membrane, 15, 19–21, 72, 73, 80, 83, 85, 90, 91, 106, 140, 153, 154, 157, 162, 163, 183, 184, 189, 192
Menten, M., 6
Mesh, 62, 97, 100, 101
Messenger, 16, 77, 118, 184
Messenger RNA (mRNA), 16, 72, 75–77, 98, 105, 106, 117–121, 129, 146, 206
Metabolic, 5, 12, 14, 18, 26, 27, 29, 30, 48, 50, 98, 175, 179
Metabolism, 2, 4, 6, 14, 27, 47–59, 71, 98, 159–180
Metabolite, 29, 73, 89, 160, 161, 185
Metal, 9, 40, 41, 49
Methionine, 56, 77
Metric, 17, 18, 22
Michaelis, L., 5–7, 9
Micromolar, 7, 29, 83
Micromole, 7
Micron, 17–19, 21
Microscope, 17, 18, 20, 185, 201
Millimolar, 7, 29, 83
Millimole, 7
Minerals, 1–16, 37, 41, 179
Misalign, 126, 139
Mismatch, 131
Mitochondria, 20, 21, 58, 67, 81–83, 85, 86, 88, 89, 91, 167
Mitochondrion, 20, 58, 73, 81
Model, 107–109, 114
Modification, 13, 28, 49, 58, 71, 72, 162, 164
Module, 61, 69, 70

Molarity, 7
Mole, 7, 27
Molecular, 1, 13, 27, 28, 30, 34, 50–51, 55, 69, 105, 107–111, 116–118
Molecular biology, 107–111, 116–118
Molecular switch, 27, 28
Molecule, 1–7, 10–12, 14–16, 18, 20, 26, 27, 29–33, 35, 38, 40–43, 45, 47–50, 52–58, 80, 81, 83–85, 87–91, 95, 96, 99–101, 103, 106, 115, 116, 118, 122, 123, 138–140, 150, 153, 161–164, 166, 173, 175, 178, 179, 184, 188, 206
Monod, J., 96
Monomer, 4
Moon, 39, 57
Mouse, 111
Mt, 81, 147
Multicellular, 20, 42, 43, 200, 201
Multimer, 69
Muscle, 16, 21, 62, 81, 83, 88, 90, 98, 116, 130, 134, 142, 162–167, 169, 171, 173, 174, 177, 179, 180, 203
Muscular, 28, 91
Mus musculus, 111
Mutation, 2, 15, 18, 33, 43, 45, 46, 56, 57, 66, 106, 124, 126–134, 138, 146, 151–155, 187, 189, 193, 194, 198, 201, 203, 205, 206, 208, 209
Mycoplasma genitalium, 14, 111
Myelin, 49
Myoglobin, 16, 81, 83, 85, 86, 88, 94
Myosin, 14, 15, 62, 134

Natural selection, 127, 149
Neuroglobin, 81
Neuron, 49, 134

Ngb, 81, 83
Nicotinamide adenine dinucleotide phosphate (NADPH), 42, 49
Nitrogen, 12, 43, 77
Nobel Prize, 107
Nomenclature, 5, 13, 33, 100, 119, 143, 163, 174, 175
Nonenzymatic, 26, 56, 161
Nonfunctional, 109, 187, 209
Nova, 40, 150
Nuclease, 20, 31, 32, 118
Nucleic acid, 22, 23, 29, 32, 33, 44, 50, 59, 105–135, 185
Nucleolus, 20
Nucleotidase, 12
Nucleotide, 12, 13, 22, 23, 29, 33, 34, 44, 50, 74, 75, 103, 104, 108, 112, 113, 116–118, 124, 129–131, 139, 163, 203, 208
Nucleus, 19–21, 42, 57, 72, 75, 91, 99, 117–119, 132

O_2, 15, 27, 43, 80, 81, 83–85, 94, 95, 152, 154
Oceans, 38, 39, 42, 53, 54
Offspring, 33, 46, 48, 130, 134
Oligomer, 83
Onc, 13
Oncogene, 208–209
1-letter, 65, 66
Orange juice, 45
Orbit, 39, 57, 149
Organ, 14, 20, 62, 80, 81, 161, 163, 165, 167, 193, 200
Organelle, 18–21, 48, 72
Organic, 30, 32, 44, 53, 58, 59
Organism, 1, 2, 14, 15, 20, 22, 33, 39, 40, 42, 43, 45, 47, 48, 57, 64, 66, 106, 110, 127, 132, 139, 144, 182–184, 200–202
Orientation, 44, 69, 74

Origin, 42, 43, 57, 114–116, 142, 163, 167, 203
Original, 4, 5, 9, 11, 45, 48, 71, 96, 108, 114, 120, 122, 139, 143, 146, 151, 155, 167, 170, 184, 189
Originate, 187
Oryza sativa, 111
Osmotic, 48
Ostrich, 22
Ostwald, W., 5
Ovum, 21
Oxidation, 48, 81, 90
Oxygen, 1–16, 21, 25–27, 33, 38, 41–44, 48, 49, 75, 77, 80–91, 94–96, 98, 144, 152, 156, 167, 174, 198, 204
Ozone, 43, 144

p21, 13
p50, 83–87, 89–91, 94, 95, 97, 156
p53, 13, 203, 206–209
Parent, 20, 106, 134, 189, 198, 205
Parental, 112, 138, 139
Pathogen, 14, 186, 191
Payen, A., 4, 5
Pentose, 108
People, 2, 6, 7, 11, 16, 17, 25, 32, 41, 45, 51, 58, 83, 107, 124, 125, 130, 143, 146, 148, 149, 154, 155, 160, 166, 171, 176, 177, 181, 188, 189, 191, 193, 198, 210
Peptide, 15, 72, 75, 77, 129, 192
Peroxide, 27
Persoz, 4, 5
pH, 6–7, 75, 88, 90, 157, 172, 175
Pharmacologist, 30
Phenylalanine, 56
Philosopher, 1, 2
Phosphatase, 103, 162, 164, 165, 168

Phosphate, 12, 13, 33, 42, 71, 103, 104, 108, 163, 164, 170, 174
Phosphofructokinase, 69, 168, 172, 173
Phosphoprotein, 13
Phosphorylase, 28, 69, 161, 163, 164
Phosphorylation, 71, 73, 103, 162–165, 175, 180, 205
Phosphotransfer, 12
Photocopy, 16
Photolyase, 27
Photons, 26, 27, 42, 48, 49
Photosynthesis, 48
Photosynthetic, 42, 49
Physicists, 37
Physics, 107
Physiological, 14, 28, 30, 75, 98, 119, 139, 151
Pili, 18
Pizza, 81
Placenta, 62, 89, 90
Planet, 37–40, 48, 53, 56–58
Plants, 1, 49, 147, 161, 166
Platelet, 20, 21, 81, 82, 97, 100, 102
Polarimeter, 4
Polarization, 4
Polygamy, 46
Polymer, 4, 20, 29, 44, 45, 50, 145, 153, 161, 164, 166
Polymerase, 112–121, 126, 127, 129–131, 135, 183, 184, 187, 188, 191, 204, 205
Polypeptide, 72, 77
Polysaccharides, 50, 166, 185
Population, 85, 124, 125, 143, 148, 182, 187, 189, 193, 194, 201
Pore, 73
Porphyrin, 48, 49
pp60, 13
Precursor, 2, 29, 33, 44–46, 50, 100, 101, 148, 149, 160, 177, 192
Primary structure, 63
Primase, 113

Primer, 113, 115, 116
Process, 4, 9, 11, 14, 16, 25, 28, 29, 32, 40, 48, 50, 57, 65, 69, 71, 73, 75, 85, 90, 98, 99, 102, 109, 111, 116, 118, 120, 132, 134, 139, 141, 142, 146, 148, 152, 153, 164, 165, 167, 174–180, 182, 184, 186, 188, 191, 193, 202, 207
Product, 2, 4, 6, 7, 10, 13, 27, 29, 42, 48, 51–53, 55, 81, 90, 91, 120, 144, 145, 166, 170, 174, 175
Proline, 56
Protease, 5, 73, 100–102
Protein, 8, 13, 15, 16, 20, 21, 29, 44, 61–77, 79, 82–86, 89, 93–100, 102–104, 106, 107, 109, 110, 117–124, 127–129, 134, 135, 138, 140, 142, 146, 151, 153, 157, 162, 163, 165, 168, 170, 172, 174, 177, 179, 181, 182, 184, 185, 188, 189, 191–193, 202–208
Prothrombin, 100, 101
Proton, 81, 90, 170, 174, 175
Pyrrole, 48
Pyrrolysine, 44
Pyruvate kinase, 62, 173–174

Q

Quaternary structure, 68

R

Radiation, 25, 43, 106, 127, 130, 143, 149–150
Radical, 25–27
Ras, 13, 103, 104, 208
Ray, 25, 43, 143, 148–150
Reaction, 2–6, 12–15, 18, 26, 30–32, 45, 52, 54–56, 66, 132, 143, 167, 171, 173–175

Reading frame, 69, 129, 170, 204, 205
Recycle, 29, 50–51, 54, 73, 178
Red blood cells, 16, 21, 41, 49, 81, 82, 87, 101, 152–154, 165, 167
Regulation, 13, 15, 117, 130, 134–135
Regulatory, 14, 15, 27, 98, 116–118, 120, 127, 135, 145, 146, 162–164, 170, 172–176, 197–210
Release factor, 77
Repair, 113, 126–134, 139, 140, 206
Replicate, 111, 184, 186
Replication, 2, 48, 111–116, 120, 129, 184, 185, 202
Research, 5, 107, 108, 200
Residue, 10, 44, 55, 100, 155, 164
Rest, 51, 80, 96, 108, 163, 167, 171
Restriction enzyme, 27, 28, 121, 124, 139
Reticulum, 72
Reverse transcriptase, 111, 191
Ribonuclease, 118
Ribonucleic acid (RNA), 6, 16, 18, 20, 22, 33, 35, 45, 57, 75, 77, 108, 109, 113, 116–121, 135, 145, 184, 185, 188, 189, 191, 193, 204, 207, 209, 210
Ribose, 33, 108, 131
Ribosome, 18, 20, 71, 75–77, 120, 129, 188, 191
Rice, 111, 148
Richardson, J., 62–64
Robot, 2, 10
Roman, 100
Rotation, 4, 39
Running, 25, 143

S

Sac, 21, 184
Saccharomyces cerevisiae, 111
Scheme, 11, 50, 80

Scientist, 4–6, 8, 9, 26, 30, 57, 63, 71, 84, 91, 107, 116, 120, 174, 207
Sea, 39, 53, 83, 84, 147, 151, 161
Signal peptide, 72
Signal sequence, 72, 73, 142
Skeleton, 62
Skin, 21, 32, 43, 62, 102, 122, 130, 142–144, 146–149, 184, 199, 209
Solar system, 37, 39, 40
Solution, 4, 6, 7, 30, 33, 83, 85, 98, 123, 152–154, 161, 184
Species, 2, 18, 22, 25, 28, 45, 57, 58, 64–66, 68, 108, 110, 113, 114, 121, 126, 129, 132–134, 138, 147, 175, 182, 184, 185, 187, 204
Speed, 10, 14, 27, 28, 98, 104, 121, 146, 208, 332
Sperm, 33, 109, 127–130
Spinal fluid, 41
Spinal nerve, 21
Sponges, 42
Src, 13
Staphylococcus, 31, 32
Star, 40, 58
Starch, 4, 5, 161
Steam engines, 51
Streptococcus pneumoniae, 111
Structural, 14–16, 57, 61, 62, 64, 65, 100, 101, 161
Structure, 12, 33, 38, 42, 44, 45, 48–50, 55, 56, 59, 61–77, 80, 83, 84, 96, 107, 108, 127, 128, 140, 143, 145, 148, 167, 172, 193
Subcellular, 48
Subset, 26, 126, 140, 153, 200
Substrate, 3–13, 27, 29–30, 32, 53, 55, 69, 71, 83, 84, 96–98, 100, 101, 143–146, 163, 170, 172–174

Subunit, 15, 68, 69, 83, 87, 95, 96, 153, 155, 162–165, 172, 173
Sucrose, 4, 166
Sugar, 4, 5, 11, 13, 28–30, 32, 48–50, 59, 71, 89–91, 131, 159–180, 183, 185
Sun, 37–40, 48–50, 57
Sunlight, 1, 53, 54, 127, 143, 147
Superficial, 102
Superoxide dismutase, 26, 27
Surface, 21, 38, 39, 54, 66, 128, 140, 153, 155, 182, 184, 187–189
Survival, 30, 126, 139, 143, 146, 149, 151, 176, 189, 191
Swann, T., 5
Swimming, 25
Symbiosis, 58
Synthesis, 2, 20, 27, 45, 49, 50, 56, 61–77, 118, 130, 145–149, 167, 184, 202, 204

T

$t_{1/2}$, 32, 35, 57
Taut, 96
Temperature, 30, 39, 89, 90
Template, 16, 112, 115, 117, 120, 126, 131
Tendon, 16, 62
Terminus, 71–73, 140
Tertiary structure, 64, 71, 96
Tetramer, 69, 87, 94–96, 153, 155, 173
Thermodynamics, 51–58
3-letter, 13, 31, 65
Threonine, 56
Thrombin, 100, 101
Thymine, 108, 119
Tides, 39
Time clocks, 103

Tissue, 9, 13, 14, 20, 21, 49, 62, 80, 81, 83, 87, 90, 91, 94, 96, 98, 101, 106, 118, 130, 134, 150, 154, 162, 165–167, 169–171, 177, 198–200, 202, 203, 207, 209
Tissue factor, 101
Torr, 83–85, 87, 90, 94, 95, 156, 157
Torricelli, 84
Toxic, 26, 56, 130, 152, 155
Traffic signal, 72
Transcription, 15, 108, 117–119, 121, 135, 149, 157, 184, 204–206
Transferrin, 15
Transfer RNA (tRNA), 75–77, 117, 120
Transfusion, 189
Transglutaminase, 100
Transition, 52–54, 95, 203
Transition state, 53, 54
Translation, 75, 77, 117, 118, 121, 145, 205
Transport, 41, 73, 81, 88, 91
Transporter, 15, 72, 73, 94, 95, 163, 165, 166
Transposon, 110, 139
Tricarboxylic, 175
Trinucleotide, 109
Tripeptide, 77
Troponin, 14
Truck, 9, 10
Tub, 9
Tube, 32, 82, 83, 174
Tubulin, 15
Turnover signal, 73
Tyrosine, 55, 56, 144

U

Ubiquitin, 73
Ultraviolet, 25, 43, 149, 150
Uncatalyzed, 30–32
Unfolded, 64
Universe, 26, 37–40
Unload, 85, 87–88, 90, 91, 94, 95
Unsatisfactory, 99
Unspecific, 100
Unstable, 26, 30, 52
Uphill, 52, 54
Uracil, 108, 119
Urogenital, 14
UV light, 27, 43, 143, 144, 148

V

Vacuole, 20
Valine, 56, 128, 151, 153, 155
Vampire, 101
Vapor, 38, 53, 54
Veins, 80, 81, 102, 154
Velocity, 6–8, 10, 97, 98, 101
Vena cava, 80
Venus, 38
Vessels, 30, 80, 81, 102, 192
Viral, 22, 35, 182–186, 188, 191, 193, 209, 210
Virus, 1, 19, 22, 23, 28, 33, 35, 82, 109, 111, 127, 181–194, 209–210
Vitamin, 45, 49, 56, 147–149
Vitamin B12, 49
Vitamin C, 45, 56
V_{max}, 6, 8–10, 29, 97, 98, 101, 102
Volume, 21, 41, 82, 91, 147
V-type, 97–103

W

Waste, 48, 81
Water, 1–16, 20, 26, 30, 32, 37–39, 41, 42, 48, 51–54, 57, 66, 69, 75, 82, 106, 161, 163, 174, 179, 183
Watson, J., 107, 108
Weapons, 102
Website, 15

Weed, 111
Weight, 7, 13, 22, 30, 55, 69, 81, 133, 160, 179
Wine, 11
Wives, 45, 46
Wolfenden, R., 30, 31
Work, 2, 5, 6, 10, 18, 29, 48, 51, 52, 85, 91, 108, 111, 114, 129, 174
Wound, 97, 100–103

Yeast, 4, 5, 11, 67, 68, 111
Yolk, 22

Zero, 27, 98, 153, 186
ZIP code, 72
Zoology, 107

GPSR Compliance

The European Union's (EU) General Product Safety Regulation (GPSR) is a set of rules that requires consumer products to be safe and our obligations to ensure this.

If you have any concerns about our products, you can contact us on

ProductSafety@springernature.com

In case Publisher is established outside the EU, the EU authorized representative is:

Springer Nature Customer Service Center GmbH
Europaplatz 3
69115 Heidelberg, Germany

www.ingramcontent.com/pod-product-compliance
Lightning Source LLC
LaVergne TN
LVHW011000250326
834688LV00003B/33